D0758714

The Pivot

The Pivot

ADDRESSING GLOBAL PROBLEMS THROUGH LOCAL ACTION

Steve Hamm

Columbia University Press
Publishers Since 1893
New York Chichester, West Sussex
cup.columbia.edu

Library of Congress Cataloging-in-Publication Data
Names: Hamm, Steve, author.
Title: The pivot : addressing global problems through local action / Steve Hamm.
Description: New York : Columbia University Press, [2021] | Includes bibliographical references.
Identifiers: LCCN 2021017237 (print) | LCCN 2021017238 (ebook) | ISBN 9780231200905 (hardback) | ISBN 9780231553834 (ebook)
Subjects: LCSH: Social problems. | Social action. | Social change.
Classification: LCC HN18.3 .H3375 2021 (print) | LCC HN18.3 (ebook) | DDC 306—dc23
LC record available at https://lccn.loc.gov/2021017237
LC ebook record available at https://lccn.loc.gov/2021017238

∞

Columbia University Press books are printed on permanent and durable acid-free paper.
Printed in the United States of America

Cover image: © Tom Rossiter, *What if We Turned Out the Lights?*
Cover design: Lisa Hamm

Tom Rossiter, a Pivot Projects participant, created the composite cover image, *What if We Turned Out the Lights?* He has a view of the Chicago skyline from the roof of his house. He imagined that, if all of the lights in Chicago were turned out, he would see the Milky Way.

This book is dedicated to Alan Dean, one of the volunteer leaders of Pivot Projects, who was passionate about young people, education, and the environment. He embodied author E. M. Forster's epigraph for *Howards End*: "Only connect . . . " He died on April 23, 2021.

Contents

Acknowledgments

I would like to thank all of the participants in Pivot Projects, in particular the leaders, Peter Head, Colin Harrison, Rick Robinson, Andre Head, Stephen Passmore, Shulamit Morris-Evans, and James Green. They and the other participants welcomed me into the group, spoke openly and honestly, and helped correct my mistakes. I believe, as Margaret Mead is quoted as saying, that small groups of committed citizens can change the world. I hope that the Pivot Projects group and the individuals in it will fulfill their dreams. I would also like to thank my editors, Myles Thompson and Brian Smith, and the book designer, Lisa Hamm.

Introduction

In the late 1950s, my family lived in a tidy ranch house in Leawood, Kansas, a suburb south of Kansas City. At the time, Leawood was in the zone where suburban streets began to give way to seemingly endless cornfields stretching to the horizon. We had a small stream running through our backyard. My friends and I spent hours playing back there.

I remember when I began to notice deaths of wild creatures. At first, I would discover lifeless little bird bodies on the grass near the stream. Then there were more dead things: rabbits, mice, a turtle. At some point we started burying the dead animals. We had a little wildlife graveyard.

Satellite view of the former Hamm property in Leawood, Kansas

Google Maps

We didn't know it at the time, but we were experiencing something that became known later as the "silent spring"—not as a metaphor but as a vivid fact of life. *Silent Spring*, of course, was the book published in 1962 by biologist Rachel Carson. It documented the adverse environmental effects caused by the indiscriminate use of synthetic pesticides, including dichlorodiphenyltrichloroethane, commonly known as DDT. It seems likely that DDT, or one of its chemical cousins, had been killing wild things in our backyard.

While we lived the American dream in our suburban ranch house, wild animals nearby were experiencing the American nightmare.

Carson accused the chemical industry of misleading the public about the risks posed by its products and the government of turning a blind eye to problems. The book was met with a storm of criticism and denial from the chemical industry and its defenders.

In spite of the blowback, *Silent Spring* became a pivot point in environmental history. The book awakened people to the dangers of pesticide use in agriculture. It led to the creation of the U.S. Environmental Protection Agency (EPA) and to laws and regulations in the United States and elsewhere designed to protect the environment and our species. Those protections included a global ban on the use of DDT except under extreme circumstances.

A Pivot Point

Now we're at yet another potential pivot point. COVID-19 and its variants started spreading rapidly worldwide in 2020, sickening tens of millions of people, killing more than 1 million, bringing the global economy to a screeching halt, and provoking civil unrest. Angry men carrying assault rifles threatened the lives of American politicians in and around state capitol buildings. In the United Kingdom, Prime Minister Boris Johnson, after at first making light of the virus, nearly died after contracting it. This has been the worst infectious disease outbreak since the flu epidemic of 1918.

The COVID-19 crisis is a wakeup call. It smacks us with the realization that we face serious risks from a wide variety of natural disasters, including viruses, earthquakes, hurricanes, wildfires, droughts, and sea level rise. Many of these threats are associated with climate change. We are not prepared to deal with them; even worse is that the way we structure society and live our lives make us more vulnerable to these calamities and tend to make them worse. In addition, we poison ourselves and the planet with pollutants and eat unhealthy food. It's like we're daring nature to take revenge on us for all of the crap we have given it since *Homo sapiens* began wiping out other animal species starting thousands of years ago.

Now we threaten ourselves with our own extinction—or, if that prospect seems farfetched, we at least face a not-too-distant future where the human condition, for a vast majority of humans, is far worse than it is today and billions of people live in abject misery. If this happens, it will be our own fault.

COVID-19 devastates us. Like climate change, it's a catastrophe of global proportions. But it also creates an opportunity. COVID-19 has taught us that all humans are deeply connected, and we share the same fate. If we join forces, act now, and act boldly, we have an opportunity to make fundamental investments and changes in the way we operate as a species that might slow the advance of global warming and otherwise improve the prospects for future generations.

In a way, COVID-19 is similar to the catharsis astronauts experience as they traverse the near-reaches of the universe in their spacecraft. They peer out of tiny portholes and observe with awe that blue, cloud-swept orb in the distance: the planet Earth. It's a reminder of how small and insignificant we are in the great scheme of things and that we have much more in common with one another than the differences that divide us. The inimitable Carl Sagan summoned such feelings in his 1994 book, *Pale Blue Dot*, which was inspired by a photograph of Earth taken by the *Voyager 1* spacecraft on its way out of the solar system. Sagan wrote: "There is perhaps no better demonstration of the folly of human conceits than this distant image of our tiny world. To me, it underscores our responsibility to deal more kindly with one another, and to preserve and cherish the pale blue dot, the only home we've ever known."

The message of the *Pale Blue Dot* is even clearer and more urgent today than it was twenty-six years ago when Sagan's book was published. We are in crisis. We must preserve and cherish our planet and each other.

Actually, we are beset by four global crises hitting at once. In addition to COVID-19 and climate change, many countries are struggling to come to terms with the toxic effects of racism, and almost every country faces intense stresses related to economic inequities. As the great Irish poet and playwright W. B. Yeats wrote in his poem, *The Second Coming:* "Things fall apart. The center cannot hold." The four crises are closely interrelated. We have to solve all of them to solve any of them.

Reconnecting

I was finishing my work on a previous book when the cluster of crises took hold. I was thinking about what to do next when the opportunity to develop this new book fell from the sky. Sitting at my computer on April 13, 2020, at our apartment in New Haven, Connecticut, I received an alert from LinkedIn that a friend, Colin Harrison, had started a new role—that of cofounder and leader at "Post-COVID-19 Scenarios Study." I knew Colin from my years working at IBM as a corporate storyteller. I had joined the company in 2009. The big attraction was IBM's Smarter Planet initiative. The company had assembled a portfolio of technologies and was helping cities, regions, and private companies use data to operate more efficiently. Colin was one of the thought leaders behind the Smarter Planet strategy. The LinkedIn alert was intriguing. I pinged Colin, and he quickly pinged back.

Colin lives in Brookfield, Connecticut, an hour drive from my apartment, but we had not been in touch. We had both left IBM several years earlier. Colin told me via email that just a few weeks previously it had struck him that the COVID-19 crisis might induce people all over the world to rethink the way we live and that governments might be willing to put a substantial slice of the money they invested in recovery into doing things differently rather than investing simply in returning to what people think of as normal. Normal, he figured, is a path to planetary ruin.

Colin had dashed off an email on March 20 to a dozen associates proposing a study of the connections between the COVID-19 phenomenon and the critical natural and humanmade systems that contribute most directly to the sustainability of life on the planet. The results of the study, he figured, might help convince government policymakers to change direction and invest to create a more sustainable society.

Several of the people to whom Colin sent his email reacted positively. One of them was Peter Head, chair of Ecological Sequestration Trust, a nonprofit organization in the United Kingdom that focuses on improving the sustainability of cities. Peter is a famous bridge builder and environmentalist. Colin, Peter and a handful of others decided to organize a global virtual collaboration of engineers, scientists, city planners, faith leaders, educators, and others to do the research Colin had suggested. The goal of their study would be to understand how COVID-19 was affecting society and to learn lessons from the crisis that would help humanity become more resilient to future pandemics while also putting it on a path to achieve sustainability goals. The group would be a new sort of social enterprise—using business and scientific techniques to solve society's problems without seeking a financial return.

Peter, Colin, and others mapped out a way forward. They would use a portfolio of technology and thinking tools to communicate, gather information, and analyze it. Experts assembled in small teams would take on twenty or more topics and generate hypotheses, test them, and fashion proposals. The group would form alliances with leaders in cities or regions around the world and work with them to apply their new thinking through demonstration projects to help make those places more sustainable and resilient. Essentially, the project would be a solutions factory.

They would also develop and present proposals to a number of policymaking bodies, including the G20, an international forum for governments and central bankers from nineteen countries and the European Union; and COP26, the United Nations Climate Change Conference, which was originally to be held in Glasgow, Scotland, in late 2020 but was pushed into 2021 by the COVID-19 crisis.

Grand Avenue swing bridge, New Haven, Connecticut

Steve Hamm

Weeks into the initiative, its leaders decided to give it a name: Pivot Projects. The reasoning was clear. Things had fallen apart. It was time to pivot in a new direction. During a Zoom discussion, one of the participants likened the situation to a swing bridge that normally carries auto or train traffic but pivots to let boat traffic pass by. He said: "What if the bridge swings back but points off in a different direction?" That's the kind of pivot this group was looking for.

Answering the Call

An amazing collection of people came together around this initiative. They were all volunteers—doing nature's work in addition to their day jobs. Nothing they suggested or did was motivated by a desire for financial gain. While many of the participants had footholds in academia, most were engineers and business people rather than scholarly researchers. They were believers in multidisciplinary approaches to problem solving. A number of them had dedicated much of their professional lives to making the world work better. Some saw this project as their best chance to move the needle on sustainability—and perhaps it would be the last one because the window seems to be closing on the opportunity to reduce global warming and avoid the worst ravages of climate change.

Others are young people focused on braking the human bus before it heads off the environmental cliff. One of those, Shulamit Morris-Evans, had earlier glued herself to the turnstiles in an office building in London to protest the environmental damage caused by fossil fuels. Another, Manish Sah, a student in Nepal, hoped to bring one of the demonstration projects to his impoverished country—teaching young Nepali farmers to practice regenerative agriculture.

The group took on some of the biggest and most complex scientific, health, and environmental issues in the world, but they also thought deeper. In the early days of the COVID-19 crisis, many people asked anew and with increased urgency about the meaning of life. Do we have some higher purpose as individuals and as a species? How should we live? What do we owe to future generations? The Pivot Projects group decided early that it would begin by asking the essential questions—of meaning and spirituality and justice and purpose—and would consider them in each of the more than twenty workstreams.

They also wrestled with some of the daunting moral and ethical challenges that emerge whenever society contemplates a significant change in direction. With nearly every major move in society, there are winners and losers. If you don't burn coal anymore, miners are put out of work and mining communities are devastated. If people choose to behave more considerately to one another and to future generations, they may have to give up freedoms and perhaps personal wealth today. How do we find a fair balance between the interests of the individual and the many, between today's humans and those in the future?

Pivot Projects was an audacious scheme. Nobody had ever tried this before. A group of several hundred people from around the world would connect virtually, agree on a way forward, and work collaborative and unselfishly—with the goal of producing novel yet practical recommendations for changing society. They'd engage with cities and regions around the world and attempt to get new initiatives launched even as the devastation of COVID-19 continued to sicken and kill people and hobble economies. And all of this would happen in a matter of a few months.

In fact, if today's collaboration and data analytics technologies didn't exist, this quest would have been impossible. In a matter of weeks, the Pivot Projects team stitched together a collection of powerful technologies—including collaboration applications and artificial intelligence tools that are on the cutting edge of what's possible.

They created a process on the fly that was built on a foundation of years of experience in getting things done. And they reacted flexibly when barriers blocked the path forward. Perhaps their work will emerge as a model for using technology to manage quick, multidisciplinary collaborations to take on complex and critical topics in the future.

There was no assurance that the team would produce valuable analysis, insights, and recommendations. This mashup of experts was making things up as it went along. Some of the early meetings were chaotic. There was an impulse to try to do too much all at once.

At the same time, there was no assurance that the world's leaders would listen. While United Nations and Western European government leaders had pledged to rebuild green, the U.S. federal government spent billions of dollars propping up fossil fuel industries and took advantage of the crisis to roll back environmental safeguards.

It was a lot to get one's head around. Still, those involved agreed that it was worth a try.

The Embedded Journalist

I wanted to give it a try, too. Within minutes of learning about the project, I saw the potential of embedding myself in it—much like journalists are placed in military units during war. I would sit in on Zoom meetings, interview the participants frequently, and follow the path of this bold yet uncharted journey. Maybe there would be a book in it.

Well, it turned out there *is* a book. My goal in writing *The Pivot* was twofold. First, even though the world is awash in proposals for addressing climate change, I sensed that this process might reveal practical new ideas that would be tried on the local level and then replicated around the world. In other words, the book might be *useful.* Second, I believed that the story of the journey that this unusual group of people undertook would be *compelling*—and could yield fascinating insights about humanity and society. In a sense, their quest was a microcosm of the efforts by people around the world to make society more sustainable and resilient in the wake of COVID-19.

In this way, the book would bear a resemblance to Tracy Kidder's great work of nonfiction, *The Soul of the New Machine*, where he followed the path of a group of engineers as they designed a novel type of computer in the 1970s under intense deadline pressure. In the case of Pivot Projects, the new machine was not a device but a new way of getting things done. The "new machine" is collaborative intelligence: bringing together people with a wide variety of points of view and areas of expertise and using modern analytics tools to help address global problems through local action.

During the early months of being embedded in the project, I struggled with the principle of journalistic objectivity. As a journalist, I have written countless articles about business, information technology, innovation, and globalization. I strive to take and express objective views. But clearly this was a different animal than my usual journalistic pursuits. I am committed to the cause embraced by Pivot Projects—saving our planet and our species. During the many early meetings, I contributed ideas and feedback with the goal of helping the enterprise succeed. I became a participant-observer. After a time, as I wrapped up the first draft of the book, I became mainly a participant. I saw that, in order for progress to be made against climate change, many people would have to commit time and energy to working with others to help make the world more sustainable and resilient, and I would be one of them. That's when I realized that the book was not just the story of a remarkable group of people and their quest but it was also a call to action—urging you, dear reader, to join with others to get stuff done. I came to think of it as a do-it-yourself guide to saving our planet.

My own Pivot Projects journey took me full circle. In the beginning, I ventured out into a global initiative, borne on the digital magic carpet of Zoom, Slack, and Google Docs. But in the end, I also returned home, inspired by Pivot Projects to engage more fully in my own small city, a microcosm of the blue dot, to try to help restore a river that runs through the city and to improve life for people who live near its banks.

This book is about the beginnings of a journey, not about the destination. That's true for me and for Pivot Projects. Unlike the typical business book, this one doesn't lay out a series of bullet-point prescriptions about how to make an organization, an industry, or the world work better. Neither does the Pivot Projects. Both the book and the group are about learning, engaging, and working with others to get something done. In stable times, it's easier to make plans and then execute them. In times of chaos and uncertainty, solutions to problems have to emerge from engagement and experimentation.

How the Book Is Organized

In the coming chapters, you will read about how the Pivot Projects took shape in the midst of the COVID-19 crisis and how the participants did their work. I'll lay out their proposals—and explore the initial reactions to them. You'll meet a number of the notable leaders and participants and the young people who want to help turn their dreams of a better world into a reality. You'll see the group engage with communities and try to turn ideas into practical actions. And you'll learn about complexity, systems thinking, artificial intelligence, and resilience—the tools and concepts that underly the group's work. Between chapters, you'll read brief profiles of some of the participants. Here is how the story is organized:

Chapter 1: The Mission. How the project got started and what the founders hoped to accomplish.

Chapter 2: The Core Team. Who they are, where they came from, and the roles they played.

Chapter 3: The Scrum. How the group did its work—and how participants explored their own humanity.

Chapter 4: Struggles. The challenges they faced and how they tried to overcome them.

Chapter 5: Remapping the World. How the group made conceptual models of human-built and natural systems.

Chapter 6: The Theory of Everything. Exploring the belief that systems thinking is the key to solving the world's most complex problems.

Chapter 7: Rethinking Resilience. A fresh look at how we view and manage risk.

Chapter 8: Talking to Robots. The group tested the idea that humans and machines can accomplish things together that neither could do as well on their own.

Chapter 9: Points of Light. A new model for engaging with cities and regions.

Chapter 10: Places. How Pivot Projects engages with places around the world to help them become more sustainable, resilient, and livable.

Chapter 11: Bright Ideas. Proposals for policymakers and solutions for communities.

Chapter 12: Connecting. What did Pivot Projects accomplish in its early days, what did we learn from it, and where does it go from here?

The Return of the Raptors

During the COVID-19 lockdown in the spring of 2020, I was paddling my kayak on a small lake about five miles from my home and I saw a bald eagle perched in a tree beside the lake. The raptor sat there for a few minutes, then swooped down and scooped an inattentive fish from just under the surface of the lake. Burdened with the weight of the struggling fish in its talons, the eagle flew in concentric circles to gain altitude and perched in a tree to devour the fish. It was the first time I had seen a bald eagle up close in Connecticut.

In the middle of the twentieth century, at the time the young me was creating a wildlife graveyard alongside a creek in Kansas, the bald eagle was nearly driven to extinction in the contiguous United States.

Steve Hamm in the headwaters of the Blue River, Colorado

Author photo

The use of DDT along with loss of habitat and illegal hunting had caused the eagle population to collapse from several hundred thousand in the 1700s to less than one thousand in the 1960s, according to the American Eagle Foundation. Today, there are hundreds of thousands of bald eagles in the Lower 48 again—and plenty in Connecticut. The restoration of the species is one of the great success stories in the history of wildlife conservation.

This near-miracle happened because the people of the United States and their elected representatives recognized that the bald eagle was worth saving. Politicians established laws and regulations designed to protect the eagle and other wild things. This was part of a broader recognition that our fate is inextricably intertwined with that of other species. We saved the eagle, in part, to save ourselves.

Now we're at another potential pivot point in the relationship between humans and the rest of nature. Rachel Carson's *Silent Spring* catalyzed a great movement, environmentalism, sixty years ago. Will the COVID-19 catastrophe bring about another moment powered by enlightenment and resolve? Will humans around the globe decide to take this opportunity to reconsider how we live, to recognize that it is unsustainable, and to take action to set themselves on a more sustainable path? This is my hope. If this book plays even a tiny role in helping that pivot happen, I will be immensely gratified.

1

The Mission

April 22, 1970, marked the first Earth Day celebration in the United States. On that remarkable day, millions of people took to the streets and public spaces to pledge allegiance to a newly minted cause: environmentalism. "Things as we know them are falling apart. A whole society is realizing it must drastically change course," Dennis Hayes, one of the organizers, told a crowd in Washington, DC.[1] Within months, the government established the Environmental Protection Agency (EPA) and passed the Clean Air Act. It was the hopeful beginning for what was to become a global environmental movement.

On the fiftieth anniversary of Earth Day, which had been renamed International Mother Earth Day by the United Nations, Peter Head took a coffee break in the backyard garden of his home in Beckenham, a suburb of London. Peter was the chief executive officer of Ecological Sequestration Trust, a nonprofit organization in the United Kingdom focused on environmental sustainability. He had been pressing for a more sustainable society for decades—first as a bridge builder; then as a city planner; and then, starting in 2011, as the head of the trust.

Peter was a respected leader in the sustainability field. He contributed to the development of the United Nation's Sustainable Development Goals (SDGs) and had authored or coauthored a number of high-profile policy reports laying out the case for sustainability. For his engineering and climate change work, he had been awarded the Commander of the Most Excellent Order of the British Empire by Queen Elizabeth II in

2011. For his sustainability work, he had been named by the *Guardian* newspaper as one of fifty people who could "save the planet," and he was cited by *Time* magazine as one of "30 global eco-heroes."

In his backyard, he contemplated some of the small wonders of nature. The garden was in full bloom: azaleas, rhododendrons, camellias, bluebells, and more. It was full of birds as well. While enjoying his coffee, Peter saw a fledgling nuthatch fly up onto a branch of a large silver birch and then perch there, flapping its wings uncertainly, and then fly off. It was a tiny thing with a black head, gray wings, and white chest. "I spent the day bathed in birdsong. It helps a lot," he remarked later.

Peter needed soothing because he was deeply disturbed at the state of the world. Fifty years after the first Earth Day, things were still falling apart—only more so. COVID-19 was ravaging a world already

Peter Head's garden

Peter Head

beset by climate change. That's why, just a few weeks earlier, Peter and a handful of colleagues had organized a global collaborative research initiative—initially the Post-COVID-19 Research Study, later named Pivot Projects—aimed at leveraging the sense of urgency generated by COVID-19 to help set society on a more sustainable and resilient path.

Peter had hatched the plan along with Colin Harrison, a retired IBM executive living in the United States who was an expert in the smart-cities domain, and Rick Robinson, a highly regarded city planner in London. The idea was to assemble a group of experts in a wide variety of fields and put them to work building ontological models of some of the primary natural and human systems that determine how the world works. They would feed the models into an artificial intelligence (AI) computing system, ask questions of the system, and then work collaboratively with the AI to develop practical solutions. The plan was to craft proposals for international policymakers and to work with cities and regions to implement some of the ideas. The founders called the local engagements "collaboratories." They would be designed to help address global problems through local actions.

Peter saw the initiative as a model for problem solving in the twenty-first century. He believed it could signal the emergence of a new type of "holistic collaborative intelligence" bringing together multidisciplinary groups of people who capitalize on advanced computing capabilities, communications technologies, and thinking techniques to solve some of the world's most complex and daunting problems. Peter wrote: "We are faced with the challenge of untangling the complexity of human and natural systems so that we can track and make sense of what is happening today in every inhabited part of the planet and predict and forestall whatever calamity is likely to happen next."

Peter has a preternaturally calm demeanor. He speaks softly, in a level tone. But he's anything but complacent. He had been driving relentlessly for nearly a decade at Ecological Sequestration Trust to help achieve an audacious goal: to save the planet. Even at seventy-three years old, he was not slowing down. Far from retired or retiring, and in spite of a host of frustrations and setbacks, he has not lost any of his resolve to set society on a new course.

The state of planetary affairs was alarming. The previous year had been the second-warmest on record, with the Earth's global average surface temperature logging in at 1.71°F (0.95°C) above the twentieth-century average. Further, nine of the ten warmest years on record had occurred since 2005.[2] That spring, the Siberian town of Verkhoyansk reported a record arctic temperature of 100.4°F, or 38°C. Miami Beach weather.

Much of Earth's warming is attributed to the accumulation in the atmosphere of carbon dioxide (CO_2), which absorbs and radiates heat. The growth of CO_2 in the atmosphere has been accelerating since measurements were first recorded in 1958—expanding from less than 320 parts per million in 1960 to nearly 420 parts per million in 2020.[3] Scientists are concerned that a load of 450 parts per million could set in motion calamitous changes in global climate—and push the planet beyond a point of no return where efforts to reduce the output of CO_2 would be futile.[4] Two theoretical physicists specializing in complex systems had concluded that deforestation was on track to trigger the irreversible collapse of human society within the next two to four decades, and all forests could disappear in 100 to 200 years.[5]

Then came COVID-19. By the time Peter sat in his garden that day, the coronavirus had spread from its apparent source in China to practically every spot on the globe. Millions of people had become ill, and hundreds of thousands had died. Government-ordered shutdowns had brought the global economy to a screeching halt and had put hundreds of millions of people out of work. And that was just the beginning. Things would get much worse.

Don't Screw with Mother Nature

Early on, evidence emerged that there was a direct connection between the coronavirus and environmental degradation. Researchers surmised that the virus had spread from wild animals to humans in a so-called live market in Wuhan, China. A report issued by the World Wildlife Fund made the case that key drivers for the emergence of zoonotic diseases are land-use change, expansion and intensification of agriculture, and consumption of wildlife.[6] At the same time, evidence emerged that air pollution, which was already one of the leading causes of death in the world, increased the health impacts of the coronavirus.[7] "We have created the perfect society for plagues to flourish in, with our dense cities, airplanes, pollution, destroying forests," Peter said. "We have created high-risk conditions. Now we need to do something about it."

As the virus picked up momentum, beloved naturalist Jane Goodall warned that humanity will be "finished" if we fail to overhaul our food systems in response to the coronavirus and climate change. "We have brought this on ourselves because of our absolute disrespect for animals and the environment," she said.[8]

Presented with this weight of evidence that the planet was teetering on the brink of ecological collapse, one option for Peter and other environmental crusaders might have been to throw up their hands in despair.

Scientists and naturalists had been warning against the predations of humans on the environment since Rachel Carson published her book *Silent Spring* in 1962. After the first Earth Day, many countries enacted environmental protections. As evidence accumulated that humans were a main cause of global warming, international bodies took up the cause. The United Nations' seventeen Sustainable Development Goals, codified in 2015, and intended to be achieved by the year 2030, included targets for clean water and sanitation, clean energy, climate action, sustainable communities, and sustainable consumption and production.

The Paris Agreement on climate change had seemed to be a major turning point in the tempestuous relationship between humans and nature. Signed by leaders from 189 nations, the agreement created a framework for curbing greenhouse-gas emissions, mitigation, adaptation and finance. The long-term goal is to keep the increase in global average temperature below 2°C above preindustrial levels—with 1.5°C as a preferred target.

It seemed, briefly, like the human race had mustered the strength of purpose to hold climate change at bay, but, all too soon, backsliding began. Multiple nations balked at setting aggressive emissions goals.

Then a wave of elections in democratic countries ushered in leaders who denied climate science, including those in Brazil and the United States. In June 2017, U.S. President Donald Trump announced his intention to withdraw from the Paris Agreement, and he began systematically to dismantle fifty years of regulations designed to protect the environment. In Brazil, President Jair Bolsonaro began removing protections on the Amazon rainforest. As a result, the Amazon, which has been called "the world's lungs," lost 3,769 square miles of forest cover between August 2018 and July 2019, an increase of 30 percent over the previous twelve-month period.[9]

When the coronavirus began to destroy lives and economies, many environmentalists hoped that governments would focus economic stimulus programs on sustainable redevelopment. United Nations Secretary-General Antonio Guterres made an impassioned plea for them to do so on International Mother Earth Day. Calling the coronavirus pandemic the biggest test the world had faced since the World War II, he said there was another, deeper emergency: the climate crisis. "We must act decisively to protect our planet from both the coronavirus and the existential threat of climate disruption," he said. "The current crisis is an unprecedented wake-up call. We need to turn the recovery into a real opportunity to do things right for the future."[10] However, government leaders were slow to respond to the call for setting the world on a new path—at least initially. Of the first $12 trillion committed to coronavirus recovery by fifty nations in the first months after the coronavirus pandemic struck, less than 2 percent targeted postcarbon economic priorities such as developing renewable energy.[11]

The Pivot

This was the state of affairs on Saturday, March 20, 2020, when Peter received an email from Colin Harrison, a former business associate. They had worked together on a couple of projects when Peter was running the integrated planning practice at the global engineering firm Arup Group and Colin was a leader of the Smarter Cities initiative at

IBM. Peter had not heard from Colin for years, but this email message lit him on fire.

Colin had forwarded to Peter and fifteen others an email he had received earlier that day from Mitch Gusat, a researcher at the IBM Zurich Research Laboratory, with whom he had worked during a long career at IBM. Mitch had sent Colin a link to an article in the *Economist* magazine[12] about a research paper published by epidemiologists at Imperial College London that helped awaken governmental leaders to the dangers of the coronavirus pandemic—leading to massive shutdowns of societies around the world. Their model predicted that the virus would infect 80 percent of the British population in three to four months if countermeasures weren't taken.[13] Mitch's hopeful comment: "Science and research can change the world, indeed . . ."

Colin had been brooding about the future of society and the planet since he and his wife Lynn returned to their home in Brookfield, Connecticut, from their second home in Switzerland on February 12. It struck him that people in the United States weren't taking the virus seriously, in word or deed. He read ravenously to learn about the virus and its spread, and was convinced that the situation was grave. "I began to see the world was reaching a tipping point. It seemed like it might tip to a different state from which it might never come back," Colin later recalled. "If the world was not able to go back to its previous shape, perhaps it could be transformed."

That's what he had been thinking about when he received the email from Mitch. Could science and research foster a major shift in direction? For more than fifty years as an engineer, his most fundamental belief had been that science could be a primary driver of progress. This dark moment in history was an opportunity to engage—and to flex the muscles of human intelligence and computing technology in a profound way. He forwarded Mitch's email to a group of people with whom he had interacted when he was at IBM. They were all believers in the power of science and of collaboration. He wrote: "I believe the world will be changed in many ways following this event. It will not revert to 'business as usual.' Taking a systems perspective, many global and national systems will, at a minimum, be flipped into new modes, and at worst may not re-start at all."

Peter responded almost immediately: "I agree and it would be great to get some work done ahead of COP26. Our platform resilience.io could be quickly adapted to do such scenario testing for a chosen city which has good data."

COP26 is shorthand for the meeting of the twenty-sixth United Nations Climate Change Conference of the Parties, which at the time had been scheduled to take place in Glasgow, Scotland, in December 2020. (It was later postponed until November 2021.) The conference would be an opportunity to propose bold new solutions to many of the most influential governmental and research leaders in the world. Resilience.io was the data and systems modeling platform that Ecological Sequestration Trust's consultancy arm, Resilience Brokers, had developed over the previous few years.

Peter's terse reply belied the excitement he felt when he received Colin's email. Like Colin, he anticipated that the COVID-19 crisis could be a major disruptive event whose effects might be felt for years. He saw that one of the immediate impacts of the coronavirus had been drastic cutbacks of the use of fossil fuels. Remarkable things were happening. The skies were clearing over cities around the world. People in Delhi, India, were able to see the outlines of the Himalayas in the far distance—some for the first time in their lives. People experienced what it was like to live in a less-polluted world. With the roar of traffic quelled, they could hear birds. Knowledge workers got a taste of how much better life could be if they didn't have to commute to work each day and spend eight or ten hours slogging away in a bland office. "I saw the chance for a pivot," Peter recalled later.

But he worried that, as soon as the crisis passed, most people would rush back to the old patterns of life. The same would go for governments. In response to the crisis, they were already pouring trillions of dollars of stimulus money into restoring economies to the status quo that existed before COVID-19—even though the status quo was hostile to large swaths of the world's people and was demonstrably unsustainable from an environmental point of view. "Huge amounts of money would be deployed. Why shouldn't society use the stimulus money to drive an environmental agenda?" he said. "We could go forward, not backward."

In particular, he saw an opportunity to help the nations of the world achieve goals laid out in the 2015 Paris Agreement on climate change, which aims to achieve carbon neutrality by 2050 (the so-called net zero CO2 emissions target), and the SDGs established by the United Nations to provide a blueprint for achieving a more sustainable and just future for all people on the planet.

The Manifesto

Peter sat down at his desk in his study and drafted a seven-page manifesto with an unwieldy title: "COVID-19 Changes Everything . . . Can It Be a Positive Catalyst for Net Zero and Sustainable Development Goals?" Peter wrote that the world faced three overwhelming threats—climate change, the coronavirus pandemic, and unknown socioeconomic changes that would emerge in its aftermath. He proposed a study conducted by a network of experts from around the world using a systems approach, models already developed by Resilience Brokers, and published studies. They would use these tools to craft proposals aimed at making human life on the planet more sustainable and resilient—and less vulnerable to future pandemics. The initiative would propose solutions to thought leaders and engage with leaders in cities and regions to try the ideas.

He laid out an ambitious timeline. Mobilization and fundraising would start immediately. Volunteer participants would be recruited and assembled in April 2020. Collaboration technologies would be set up for them to use. First drafts of proposals would be circulated starting in July 2020. Final drafts would be published on November 15, 2020. The funding required was £350,000 (about $440,000). The idea was to run the experiment for nine months and then see what happened. It was a tidy plan for a process that would become complex, unpredictable, and by turns empowering, frustrating, chaotic, and gratifying.

During those early days, Colin, Peter, and Rick launched a flurry of communications via email and Zoom to firm up plans based on Peter's proposal. They all had deep backgrounds in sustainability and urban regeneration projects. They could go in a lot of different directions. Which would have the most impact? Which was most doable? After a few days of email exchanges and phone calls, Peter asked the others: "Well, should we do it?" Rick later recalled the moment. He had never done anything like this before. There was no precedent for it really. He knew the project could collapse at any point. He said: "I'm too stupid to know this won't work, so I think we should try anyway."

It was a go. They began reaching out to the vast networks of contacts they had amassed over their careers—people in government, corporations, academia, and nonprofits. Peter used his Twitter account to announce the initiative and invite people to join. One of the first people to respond was Nigel Topping, who was appointed by the UK government in January 2020 as its high-level climate action champion for the COP26 climate talks. He would be the host of the conference. That seemed like a big win.

One contact led to another. Some people immediately volunteered to participate; others watched from the sidelines. Within two weeks, more than fifty people were on board. In a month, the tally was 120. Most of the initial joiners were experts from the United Kingdom and the United States—and they tended to be older. The group needed more young blood. One of the early responders, Anh Nguyen, a Vietnamese woman who was studying in Sweden, put out a call for younger volunteers, and dozens joined. By the end of 2020, the project had attracted more than four hundred volunteers from twenty-five countries—though the truly active participants numbered fewer than one hundred at any one time.

Almost immediately, the organizers began setting up a platform for getting work done. The technologies would include Zoom, the videoconferencing system; Slack, an online collaboration platform; and computer modeling and artificial intelligence technologies. The reasoning tools would include systems thinking, the complex adaptive systems theories developed by the Santa Fe Institute in the United States, and The Cynefin (pronounced "kinevin") Framework, a method for understanding complexity developed by the Cynefin Centre in the United Kingdom.

There would be more than twenty small working groups. In the center of the organization would be the core workstreams, including energy, climate change risks, economics, and others. Clustered around them would be three other categories: climate change themes aligned with the Paris Agreement goals, societal progress themes aligned with the SDGs, and COVID-19-specific themes. They also included workstreams addressing arts and culture, education, communities, and beliefs.

Each workstream was to choose a leader and spend the first weeks completing an assignment, which included getting to know one another, identifying a handful of key questions they wanted to answer through their work, and preparing spreadsheets where they would create the system models they expected to explore and the relationships between systems. The groups would use online visualization tools to help them build and learn from their systems models. And finally they'd use AI to explore potential solutions. Rick referred to the entire process as "extreme science."

A number of tech companies had developed question-answering and hypothesis-generating AI machines. The group quickly zeroed in on one of them: SparkBeyond. When Nigel Topping responded to Peter's initial Tweet, he pointed him to the small Israeli tech company. Its founders had set out in 2013 to build an AI-powered, problem-solving technology. The system was designed to help groups of people overcome the biases inherent in human thinking, identify root causes of problems, discover patterns, and search for solutions. The company also had a social mission: to drive meaningful change on a planetary scale.

From the start of engagement, it seemed that Pivot Projects and SparkBeyond were a nearly perfect match for one another. The group could feed its models into the machine, test them, and add more data and contextual information. After the prep work, the humans could work interactively with the machine to target and explore the most promising hypotheses—teasing ideas that could support environmental sustainability and equitable growth.

Colin referred to the goal of this work as a whole-Earth model that could be used not only by Pivot Projects but by any researcher or research group that found value in it and was willing to add to it. Just as the open-source software movement had transformed the tech industry and the open-data movement had helped cities solve problems, the open-Earth-modeling movement could help solve the biggest problems of all.

The Importance of Systems Thinking

Fundamental to the way Pivot Projects was conceived, was organized, and operated was the belief among its founders in the value of systems thinking. Peter, Colin, Rick, and a number of the other key members of the enterprise believed that understanding the behaviors and interdependencies of systems was key to solving the world's problems.

Systems thinking is an approach to viewing the world as a complex system made up of subsystems, both natural and humanmade, with the goal of understanding more deeply how the world works so that people can design interventions that could improve its performance based on their values and goals. A system of interest might be a watershed or a metropolitan transportation hub. Systems succeed when they operate in balance; they fail when they are disrupted or slip out of balance.

At the core of systems theory is complexity science, which is the study of a certain kind of system—the complex adaptive ones. Complex adaptive systems are systems that have a large number of components, often called agents, that interact, adapt, and sometimes learn. Notable examples of complex adaptive systems are climate, cities, markets, industries, natural ecosystems, social networks, war, terrorist networks, the internet, the brain, and the immune system—and the universe.

Chapter 6 of this book provides a more detailed exploration of problem solving through systems thinking, but because the practice is so fundamental to Pivot Projects, I'm laying some of the groundwork here. Consider these paragraphs to be a systems-thinking starter kit.

Systems theory and complexity science are two of the most significant advances to thinking and problem solving that have emerged over the past several decades. The goal of systems thinking is to be able to model the dynamics, constraints, and conditions existing in a system so you can understand the mechanisms that make the system work the way it does. Modeling of simple systems can be done manually, but models of complex adaptive systems are built and manipulated with the help of computers. Scientists and social scientists build models of systems in software using ontologies—which define and describe the systems. Then they load them into computers, and increasingly into computers that are capable of learning. The computers can be used to run simulations on the models—making it possible to study the actions and interactions of components in the systems and to produce what-if scenarios. Let's say that you're thinking of building a highway along a river through the center of a city. What are potential effects on commuting times, water and air pollution, and the number of asthma cases in the adjoining neighborhoods?

At the highest level, systems thinking is a framework—a way of understanding how the world functions. At a more actionable level, it's a way of studying how particular systems of interest work and interact so you can intervene to make them work better. But there's also a social aspect of systems thinking: ideally, it's performed by a sizable group of people with different types of expertise and points of view.

That's quite a switch. Traditionally, thinking about hard problems is done by experts in narrow domains operating in relative isolation from one another. That may seem fine if you believe deforestation in the Brazilian Amazon is not related to desertification in North Africa and a refugee crisis in southern Italy. But, of course, we now know they *are* related. Systems thinking breaks down the silos of knowledge and puts interdisciplinary groupings of people in a room or on a Zoom call where they can learn, design, and solve problems in a holistic, collaborative way.

Pivot Projects' leaders were hyperaware of the blinders that they wore as scientists and engineers. They understood that disciplines and approaches based on science and engineering can't deal with complex situations on their own—even with the assistance of powerful tools such as data analytics software and AI. They knew they would have to take into account other factors, including politics, law, economics, community relations, religion, and history, to help find and fashion sustainable solutions to problems.

Peter Senge, director of the Center for Organizational Learning at the Massachusetts Institute of Technology's Sloan School of Management, has long argued for a fundamental change in how we think about thinking. "The smartness we need is collective. We need cities that work differently. We need industrial sectors that work differently. We need value chains and supply chains that are managed from the beginning until the end to purely produce social, ecological, and economic well-being. That is the concept of intelligence we need, and it will never be achieved by a handful of smart individuals."[14]

Integrating Beliefs and Values

A notable feature of Pivot Projects was its inclusion of spiritual life and beliefs in its deliberations. One of the workstreams, called Faiths, would endeavor, like the other groups, to map models of systems within its domain. In addition, members of the group decided to sit in on other team meetings to make sure all of them took beliefs and ethics issues into consideration when they modeled their themes. Some members of the Faiths group saw themselves as a necessary counterpoint to the science and engineering points of view that were so prevalent in other participants.

It was Peter Head who insisted on making beliefs and culture major components of the study. He had contributed to the development of the Earth Charter, an international declaration of fundamental values and principles that was rolled out in 2000. It laid out an ethical framework stating that environmental protection, human rights, equitable economic development, and peace are necessary for human progress in the twenty-first century.[15]

He later came to understand the importance of including local people and their values in city planning decisions. When he worked for Arup, he had been a member of the City of London Sustainable Development Commission. One public housing project he worked on was located in the East End of London, near the port, which had been the home for wave after wave of immigrants beginning in in the nineteenth century. The city had done extensive outreach to get input from the community. Sustainability and reducing waste were essential elements of the proposal. After a public meeting, a man had approached Peter and put his hand on his arm. He was an elderly Sikh wearing a turban, with a long white beard and flowing mustache. The man, who did not give his name, told Peter that conserving resources and avoiding wasting were key tenets of the Sikh religion, but there had been no recycling program in his neighborhood previously. He told Peter: "Now you're bringing this ethos. It means London feels like home to us for the first time." Peter described the encounter as "a powerful moment in my life." From then on, in his sustainability work, he always tried to include people from faith groups who not only brought their religious sensibilities but often spoke for the voiceless masses. "We need all of this kind of input to make what we talk about resonate with people all over the world," Peter told me shortly after Pivot Projects launched.

Jeff Newman, a retired rabbi who had served one of London's largest Jewish congregations, emerged within Pivot Projects as a spokesperson for bringing faiths and beliefs to bear on Pivot Projects. To him, it was critical to recognize the interconnectedness of people—their shared values and their empathy for one another. The engineers in the group talked about systems of systems. Beliefs, he thought, were the threads that stitched the other systems together into a cohesive whole. "The Faiths group brings the sense of oneness," he said. "It's the holistic approach that unites us all in the project."

During several video calls, members of the group said they were inspired by an op-ed published by Mark Carney, a governor of the Bank of England, in the *Economist* magazine on April 16, 2020. In it, Carney warned that the damage to the global economy wrought by COVID-19 was hard to predict and would likely be profound and long-lasting. But he saw the disaster as an opportunity to change course. He pointed to philosopher Michael Sandel's assessment that capitalism's market economy had evolved into a "market society," where every asset or activity has a price tag attached to it. If it can't be bought or sold, it is viewed as having little value. Carney called for a reversal. In the future, he said, public values should shape private values. His parting shot: "Once this war against an invisible enemy is over, our ambitions should be bolder—nothing less than to make a fit planet for our grandchildren to live on."

The group decided that this was to be a mission of Pivot Projects: to help society move from a focus on monetary value to a focus on human values.

POEM
By Kristin Flyntz

An Imagined Letter from COVID-19 to Humans

Stop. Just stop.
It is no longer a request. It is a mandate.
We will help you.
We will bring the supersonic, high-speed merry-go-round to a halt.
We will stop
the planes
the trains
the schools
the malls
the meetings
the frenetic, hurried rush of illusions and "obligations" that keep you
    from hearing our
single and shared beating heart,
the way we breathe together, in unison.
Our obligation is to each other,
as it has always been, even if, even though, you have forgotten.

We will interrupt this broadcast, the endless cacophonous broadcast
of divisions and distractions,
to bring you this long-breaking news:
We are not well.
None of us; all of us are suffering.
Last year, the firestorms that scorched the lungs of the earth
did not give you pause.
Nor the typhoons in Africa, China, Japan.
Nor the fevered climates in Japan and India.
You have not been listening.
It is hard to listen when you are so busy all the time, hustling to
uphold the comforts and conveniences that scaffold your lives.
But the foundation is giving way,
buckling under the weight of your needs and desires.
We will help you.
We will bring the firestorms to your body
We will bring the fever to your body
We will bring the burning, searing, and flooding to your lungs
that you might hear:
We are not well.
Despite what you might think or feel, we are not the enemy.
We are Messenger. We are Ally. We are a balancing force.
We are asking you:
to stop, to be still, to listen;
to move beyond your individual concerns and consider the concerns
of all;
to be with your ignorance, to find your humility, to relinquish your
thinking minds and travel deep into the mind of the heart;
to look up into the sky, streaked with fewer planes, and see it, to
notice its condition: clear, smoky, smoggy, rainy? How much do
you need it to be healthy so that you may also be healthy?
To look at a tree, and see it, to notice its condition: how does its
health contribute to the health of the sky, to the air you need to be
healthy?

To visit a river, and see it, to notice its condition: clear, clean, murky, polluted? How much do you need it to be healthy so that you may also be healthy? How does its health contribute to the health of the tree, who contributes to the health of the sky, so that you may also be healthy?

Many are afraid now.

Do not demonize your fear, and also, do not let it rule you. Instead, let it speak to you—in your stillness,

listen for its wisdom.

What might it be telling you about what is at work, at issue, at risk, beyond the threats of personal inconvenience and illness?

As the health of a tree, a river, the sky tells you about the quality of your own health, what might the quality of your health tell you about the health of the rivers, the trees, the sky, and all of us who share this planet with you?

Stop.

Notice if you are resisting.

Notice what you are resisting.

Ask why.

Stop. Just stop.

Be still.

Listen.

Ask us what we might teach you about illness and healing, about what might be required so that all may be well.

We will help you if you listen.

Author's Note: This poem by Kristin Flyntz was first published on a website, Dark Matter: Women Witnessing (http://darkmatterwomenwitnessing.com/issues/Oct2020/articles/Kristin-Flyntz_Imagined-Letter-from-COVID-19.html). It was brought to my attention by Jeff Newman, a Pivot Projects participant. Here's Kristin's explanation: "I have a confession to make: I'm not a poet. At least, not in any formal sense. 'An Imagined Letter from Covid-19 to Humans' was an unexpected gift that 'came through' one morning before work. I never, ever expected it to be shared at all—let alone as widely as it has been, and I am deeply grateful that it has resonated with so many people." As she was typing this email message, a black bear wandered through her yard in suburban West Granby, Connecticut.

2

The Core Team

Although Pivot Projects' leaders had a bold vision of what they wanted to accomplish from the start, it took a while to choose a formal name. For the first several weeks, the project was called the Post-COVID-19 Research Study, but clearly that wouldn't do. It wasn't catchy. The word *pivot* kept popping up in conversations, not just among themselves but with world leaders, pundits, and letters-to-editor writers alike. The name Pivot Project was already taken, so they decided to make their name plural: Pivot Projects, it was to be.

Over several weeks, the Pivot Projects took on a more formal leadership structure. The three initial leaders, Colin Harrison, Peter Head, and Rick Robinson, began to take on distinctive roles and were joined by Andre Head, Peter Head's son, who worked as a designer and collaboration coordinator for Resilience Brokers. He would manage the technology, onboarding of volunteers, and training. They were the core team. Each of the leaders had taken a personal journey that prepared them for their roles and fostered their determination to get something important done.

Peter's Journey

It will probably come as no surprise that Peter was a Boy Scout in his early days. He grew up in Surrey, just south of London. His father was a civil servant. This was after the war, when the British economy was still recovering. Peter remembers food rationing and watching his parents struggle to make ends meet. They had to conserve to survive. Scouting was his passion and his doorway to the world. He got to hike and camp all over England. He rose through the ranks to leadership—first as a patrol leader, then as a group leader, then as a Queen's Scout. He charts his interest in engineering to times when he built rope bridges over streams and temporary shelters on mountainsides. As a teenager, he dreamed of being an architect, but he veered into engineering studies because his father spotted an advertisement in the *Daily Telegraph* newspaper touting university scholarships for civil engineers and encouraged him to apply for one—which he landed.

Peter was the first in his family to get a college education. He graduated with a degree in civil engineering from Imperial College London and embarked on a career in the bridge-building field that lasted more than thirty years. He worked on bridges around the world. He started as an engineer but rose through the ranks to the point where he was managing huge bridge projects, from planning to completion. In midcareer, he pioneered innovations in advanced composite technology. Working with colleagues at the civil engineering firm Maunsell, he developed a system for using durable reinforced polymer materials in bridge decks that made the spans both stronger and more flexible. They later designed and built a bridge entirely from composite materials. As a project manager, he improved the process of planning bridges and managing their construction.

The highlight of his bridge-building career was his work on the Second Severn Crossing, a three-mile-long bridge that connects England and Wales. It's now called the Prince of Wales Bridge. While working for Maunsell (later acquired by AECOM), he was appointed the government agent overseeing the project in 1984. The experience is what turned him into an advocate for sustainability.

This was a particularly challenging project. The wide estuary where the Severn River flows into the Atlantic Ocean is subject to high winds, strong currents, and extreme tides. The bridge had to be able to withstand the corrosive effects of seawater and potential collisions with large ships. Much of the estuary is mud flats that are exposed at low tide. The flats had been designated by the government as a special environmental protection area because they were home to migrating birds, including the ringed plover and redshank. So the bridge had to be designed to withstand tremendous natural forces but at the same time cause minimal disruption to nature. And the construction process had to do the same. It was a systems problem. All sorts of complex interdependencies came into play. A collaborative multidisciplinary approach was required for dealing with them. At the time, there were no rules in place governing environmental impacts of construction projects. Peter and his team had to write them. After the bridge was completed in 1996, government surveys showed that it had minimal impact on the natural environment.

During this project, Peter came to understand that it was important for bridge builders to consider not just the physical structures but the impact that bridges have on nature and society. He began to formulate a new business model for bridge projects. Rather than simply requiring engineering and construction companies to complete projects on time and on budget, he developed the concept of performance-based procurement. Contracts were reframed. Builders were required to commit to delivering something of value to society—taking quality of human life and nature into account. "Everything we invest in should be seen as a service to society and not a thing," Peter said later.

Peter was on a path to a critical juncture where he would become more interested in saving the planet than he was in building things. He served in a couple of roles that broadened his horizons. One was his appointment to the London Sustainable Development Commission. The other was a visiting professor appointment at Bristol University, where he helped create C40, a network of cities committed to addressing climate change. The realization that so much of the built environment was not environmentally sustainable prompted him to switch careers and begin designing sustainable cities with Arup. At the time, there was still a lot of skepticism about climate change. The firm commissioned a report to discover whether concerns about global warming were warranted. It turns out they were.

Engineering for Sustainability

At Arup, Peter established and led an integrated planning organization incorporating planners, urban designers, and other professionals. One of his biggest projects was the Dongtan Eco-City program, one of the early initiatives in China aimed at building new cities designed from the start to be low-carbon and ecofriendly. The project brought together urban planning and sustainability experts from around the world to work with Chinese government leaders. As bold as the project's ambitions were, it soon ran into the realities of China's political economy. The cities were supposed to be ultra-efficient and resource-conserving, yet Chinese leaders faced intense pressures to produce jobs and increase gross domestic product (GDP). They were committed to using lots of resources quickly to do so. "We produced plans but the plans weren't being implemented," Peter said. Frustration set in.

Then Peter got a bully pulpit from which he hoped to change minds. The Institute of Civil Engineers, of which he was a member, chose him to prepare a report and deliver lectures to engineering groups around the world. The institute asked him to choose a topic related to sustainability. Peter took a leave of absence from Arup because this assignment gave him the excuse to sit down and think deeply—distilling the knowledge he had accumulated over the past decades as a bridge engineer and city planner into a cogent theory. He went big. The pitch was that engineers could help design a new world where humans lived in harmony with nature.

Peter had come to believe that humanity has the potential to move to a sustainable way of living over the next few decades—adapting to climate change impacts while at the same time allowing for continued economic development and population growth. His goal is for humanity to enter what he called the Ecological Age by 2050. He believes developed nations will achieve the goals by retrofitting established infrastructures and systems in addition to building new ones. In developing nations, much of the change will come through establishing new models for development rather than following the paths that developed nations took in the twentieth century.

His eighty-three-page report, "Entering the Ecological Age: The Engineer's Role," was published in 2008.[1] The report was full of facts and charts. He laid out three targets that will mean we have entered the Ecological Age:

- Reduction of carbon dioxide (CO_2) to 1990 levels.
- Reduction in ecological footprint in all countries to 1.44 global hectare per person (GHP). (GHP measures the earth's biocapacity in production land area, including cropland, pastures, forest, and fisheries.)
- A substantial increase in the UN Human Development Index based on GDP per capita, life expectancy, and education attainment levels.

The report also included detailed proposals for changes in transportation, water and waste, energy, food production and distribution, and broadband communications. He believes that all of these systems are connected and form virtuous cycles that integrate the environmental, economic, and social performance of components in the built environment so one design change can lead to benefits in another.

Peter took his ideas on the road. With financial support from the Institute, he visited twenty-eight countries over an eighteen-month period—presenting to more than sixty audiences. The experience of sharing his ideas and sense of urgency was gratifying. The presentations generated discussions and a lot of enthusiasm. They were like Al Gore's "An Inconvenient Truth" lecture series arguing for action to combat climate change—only for civil engineers.

Once again, however, he was left feeling frustrated. What more could he do? His answer came a short time later, at a climate change conference in Hong Kong in November 2010 during a presentation by Chinese and European environmental ministers. The ministers discussed the prospects for climate action by governments. Afterward, during the question-and-answer session, Peter raised his hand. When called on, he pointed out that emissions from coal-fired power stations were a major contributor to the accumulation of greenhouse gases in the atmosphere. Why not use carbon sequestration techniques to bury the carbon emissions underground? They could use a natural process to

Peter Head at the conference in Hong Kong in November 2010

Peter Head

save nature. The answer from the stage was unsatisfying: great idea, but not our job.

After the session ended, Peter grumbled: "It's always somebody else's problem to solve everything." A woman sitting one row in front of him asked: "If you had the resources, what would you do?" He answered that he would bring together a wide variety of experts to focus their minds and data on the problem, they would solve it on a small scale in one location, then they'd spread their solution around the world. She thought about his answer for a moment, then replied: "Why don't you do it, then?" That was the genesis of Ecological Sequestration Trust.

After Peter returned to the United Kingdom, he talked with his family about the prospect of starting a charity. His wife and children encouraged him to go for it. In 2011, Peter left Arup. He had just turned sixty-five. He cashed in one of his pensions as seed money to star the trust. His idea was that the trust would work with leaders around the world to help nations, cities, and regions achieve sustainability goals. Key elements of his plan included using systems thinking, building models to map the interactions between human and natural systems, and tapping into networks of experts to solve problems. He called the whole thing collaborative intelligence.

With his reputation as a planet-saver as his calling card, Peter was able to form collaborations with like-minded individuals and organizations. He convened a series of high-profile conferences and produced some cogent reports—the most significant one being "Roadmap 2030," a 136-page report laying out proposals for financing and implementing the UN's Sustainable Development Goals (SDGs) in cities and regions by 2030.[2] It included an introduction by Jeffrey Sachs, director of the Earth Institute at Columbia University. Sachs was one of the architects of the SDGs.

The trust worked on a series of projects aligned with its mission. These projects were funded by governments and foundations. The most significant of them was a project aimed at improving water quality and sanitation in Accra, Ghana. In 2016, the trust's experts used systems thinking and modeling techniques to help leaders redesign the water infrastructure. While the work was well received, the government didn't have the money to fund a major infrastructure project.

Money was a perennial problem. Peter tried to raise about $100 million from corporations and foundations to pay for a ten-year program to develop the data and modeling, and while some parties showed interest, in the end, nothing came of it. Funding organizations were project focused. They didn't want to pay for long-term planning and capacity building. So the trust forged ahead with one-off projects, operating on a shoestring budget. It had not achieved the kind of impact Peter had hoped for.

It is no wonder that Peter plunged so quickly into Pivot Projects. The enterprise was in many ways a continuation and elaboration of a mission he had been pursuing doggedly for nearly a decade—without a lot to show for it. Perhaps this was the breakthrough he had been hoping for.

Colin's Journey

Colin had spent his earliest days in Sheffield, the longtime center of steelmaking in the United Kingdom. As a small child, his interest in engineering became apparent. When he was about to turn five years old, his parents gave him a Meccano kit, one of those construction sets that enables kids to explore the principles of mechanical engineering by building structures from nuts, bolts, and metal pieces. (The counterpart in the United States was the Erector Set.) Colin couldn't read yet, but he managed to assemble things by looking at pictures in the printed directions. "That was the defining moment. I wanted to be an engineer. I never wanted to do anything else," he said.

Looking back, Colin believes that his early interest in spatial thinking gave him a head start in seeing complex relationships among things.

The family later moved to Scarborough, a resort town near the North Sea, where Colin's dad ran a restaurant equipment business. Because of steelmaking, Sheffield had been heavily polluted. As a small child, Colin actually liked the odor of the place. He thought it smelled like malt. Later he realized he was smelling the results of toxic chemical reactions. In contrast, the air was clear in Scarborough. On frequent bike rides outside the city, Colin came to appreciate nature for the first time. There were beautiful cliffs next to the sea north of Scarborough. Inland was North Yorkshire Moors National Park. A beautiful river ran through the park's Langdale Valley, and, if a boy was inclined, he could strip off his clothes and jump in a swimming hole. Colin remembered thinking: "I'd really like to get to know this valley, so I'd know every tree and rock in it."

He also recalled a practice in Scarborough that today would be called an example of a circular economy. The town was full of restaurants catering to the tourists. Lots of fish and chips and other meals were served. Entrepreneurs operated services that collected food waste from the restaurants and delivered the material to farms on the outskirts—where the scraps were fed to pigs.

Like Peter, Colin studied engineering at Imperial College London (where he discovered that he missed hearing the crash of the surf in the distance). After he got his PhD, he landed his first professional assignment as an engineer working on one of the most important scientific collaborations in the twentieth century: the twenty-three-nation European Organization for Nuclear Research, commonly called CERN. The organization has produced some of the most important discoveries in physics, including the famed Higgs boson particle. To enable easier sharing of documents and data, scientists there developed the World Wide Web. Colin began as an engineer in 1972 working in the particle physics laboratory on a component of the Super Proton Synchrotron (SPS), a two-kilometer, donut-shaped tunnel used for accelerating high-energy protons, electrons, and positrons. Colin's part was designing fast-pulsed magnets for switching the beams between different trajectories. After the SPS was finished, he ran overall engineering operations for a couple of years.

Since childhood, Colin had been especially tuned to the relationships between things—not only physical objects but also concepts, especially concepts spanning multiple disciplines. During the construction and operation of the SPS accelerator, he felt very uncomfortable until he had built up a mental model of how it all worked.

The contrast between Colin's early days at CERN with Pivot Projects is striking. Even though CERN was a collaboration, the organization and its processes were highly structured and compartmentalized. Interactions between human beings took place in person or on the phone—but conference calling didn't exist. One of Colin's colleagues once informed him that it was possible for the technical staff members to send messages to one another on the computing system. Colin asked, "Why the hell would you want to do that?"

After Colin left CERN, he was one of the engineers who designed the world's first clinical magnetic resonance imaging (MRI) machine. The job was almost entirely technical, but he relished it. He quickly became the leader of the engineering team and was entranced by the ways in which radio-frequency and magnetic-gradient pulses could induce elaborate choreographies by the weakly magnetic hydrogen atoms in human tissue, thereby permitting a wide range of diagnoses.

Years later, when Colin was at IBM, he began to organize collaborations more like Pivot Projects. In the early 2000s, IBM's leaders became

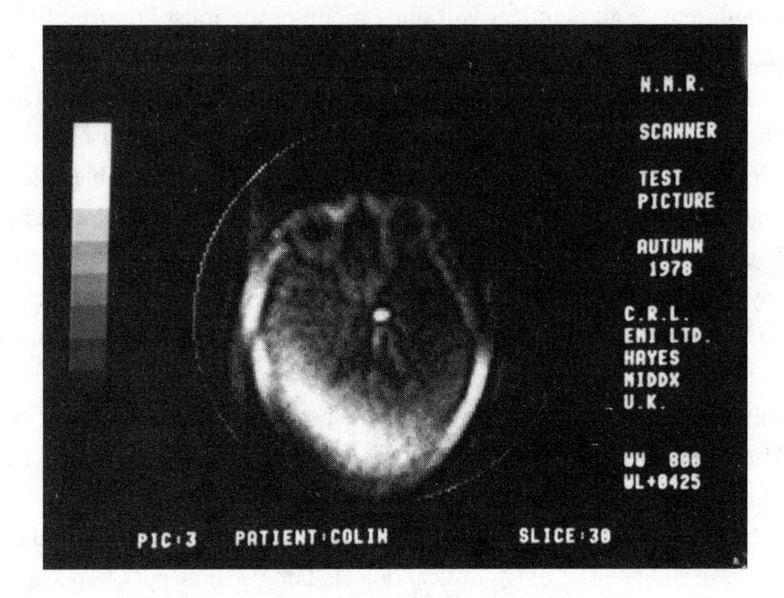

First MRI of the soft tissue of a human brain—Colin Harrison's

Colin Harrison

interested in promoting innovation—both within IBM and in collaboration with business partners and clients. This was after Harvard professor Clayton Christensen published his seminal book, *The Innovator's Dilemma*, about why it was difficult for large successful companies to produce major disruptive innovations. Colin, who worked at IBM's corporate headquarters in New York at the time, spotted an opportunity and organized an internal innovation club, ultimately comprised of about 200 IBMers from around the world. They would get on weekly conference calls and discuss their innovation projects and the latest management theories of innovation.

A few years later, IBM launched its ambitious company-wide initiative called Smarter Planet. It originated, in part, with Colin. In 2006, IBM employees participating in an online "innovation jam" had called for the company to launch green business initiatives. Colin, who worked in IBM research at the time, was asked to comb through the employee feedback to look for potential green business opportunities. He found some places where existing IBM technology could be adapted to help cities and organizations operate more efficiently and made a pitch to then-CEO Sam Palmisano. Before too long, the government of Abu Dhabi, in the United Arab Emirates, invited tech companies to provide input for its plans to create a new city in the desert, to be called Masdar City. It was to be a sustainable, mixed-use development capitalizing on the latest digital technologies.

One evening, after talking to the government planners, Colin and his IBM colleagues reconvened in a conference room in the company's offices in nearby Dubai. Somebody drew a large triangle on a white board with the word *brain* written at the apex. Along the bottom of the triangle, this person listed all of the systems that would have to be designed and built and controlled in an integrated manner to create a sustainable, high-tech city. To be effective, all of those systems would have to be integrated and coordinated and managed. "We saw not just a collection of solutions. We saw it as an 'intelligent city,' " Colin recalls. The team initially called this initiative the Intelligent Cities, but it was renamed Smarter Cities and folded into the Smarter Planet strategy.

Meanwhile, IBM's marketing department, based on their own research into the results of the innovation jam, had been thinking along the same lines. They came up with the Smarter Planet label, and Smarter Planet became a company-wide initiative.

The Smarter Planet strategy was rolled out in the middle of the global financial crisis and recession of 2008. The thought was that governments would spend aggressively to stimulate their economies, and one of their priorities would be on funding infrastructure projects. IBM and other tech companies pitched governments the idea of investing in smart infrastructure that would use networked sensors to collect massive amounts of data and use software to analyze the data so they could make better decisions and manage their resources more efficiently. This was about easing traffic jams, preparing for disasters, improving sanitation, cutting water waste, improving food safety, and the like. But they also produced and proposed Smarter Planet solutions for the private sector. Ultimately, IBM developed technologies harnessing connectivity, sensors, and analytics aimed at improving the performance of just about every human activity.

The Smarter Planet campaign immersed Colin in the domain of systems thinking. He was the technical leader of the most novel component, Smarter Cities. Many U.S. and UK cities had been in decline since the 1950s, when the middle class began to move to the suburbs. The cities were plagued by big social problems and shrinking budgets. Others were booming again, including New York City and London, but suffered from overuse. Meanwhile, around the world, the mass migration of country people to sprawling megacities was accelerating. At the time, about 50 percent of the global population lived in cities, and it was expected to increase to 69.6 percent by 2050.[3] Cities had reemerged as the world's economic engines, but their density and complexity made them hard to manage. Colin joined a fraternity of scientists and urban planners who believed the problems of cities could be understood and addressed only by viewing them as complex adaptive systems—where the whole is more complex than the sum of its parts and where the component systems interact and evolve continuously. They believed that only by building models of these complex systems of systems could technocrats understand their interactions well enough to design interventions what would improve the quality of life and efficiency of cities.

By reading voraciously and interacting frequently with city leaders and urban planners, Colin was able to build a reputation as one of the IBM executives who truly understood how cities work, and as a result he became a figurehead for the Smarter Cities campaign. He participated in dozens of engagements with cities all over the world, and he came to understand deeply the challenges in helping cities work better. He saw that these efforts had to be multidisciplinary, collaborative, data-rich, and real-world informed.

In an effort to help build these societal capabilities, Colin created a multidisciplinary community of professionals, which he called the Urban Systems Collaborative. The initiative, which was sponsored in part by IBM, brought together professionals from around the world representing an array of disciplines. They studied urban problems, engaged in debates and show-and-tell sessions, and convened for annual conferences—where they heard from leading thinkers in urban studies and related fields. It was like his IBM innovation club on steroids.

As Colin's career progressed, he became less interested in the technologies that people use to get their work done and more interested in understanding how the world works and how people can collaborate to make it work better. He retired from IBM in 2013 but kept thinking and writing—focusing on the workings of cities and on collaboration between humans and thinking machines. Colin became frustrated with the lack of an underlying science of how cities operate, and that was the itch he kept scratching.

Rick's Journey

Rick and Colin had met at IBM, where Rick also played a part in the Smarter Planet initiative. He pitched the smart-city vision to leaders in the United Kingdom. He learned early that the city leaders of Sunderland, a city on the northeast coast of England, were interested in exploring the possibilities. When he and his colleagues arrived, they were introduced to a woman named Margaret Elliot, who was founder and managing director for an organization called Sunderland Home Care Associates. It provided care for people who had difficulty taking care of themselves. She was a working-class woman who, with a small group of other women, had cofounded the organization, which was focused on delivering high-quality care at affordable prices. She explained that it was started in one of the most deprived areas of the city, which was home to both its customers and its employees. The organization not only provided care in the area, but it also created jobs and it made its employees shareholders.

At first, Rick was confused. Why were they talking to this woman rather than the city's elected leaders and bureaucrats? Then he got it. The Sunderland people were sending him and his IBM colleagues a clear message: For them, this engagement couldn't be about gee-whiz technology; it was all about figuring out how to make people's lives better, as Elliot had done. Reflecting back, Rick says: "That experience defined what I did with my career for the next thirteen years and it explains why I am involved in Pivot Projects."

Rick had grown up in Hampshire, in the south of England. His father worked for IBM, and at one point he was director of the company's famous Hursley software laboratory. Rick's mother had been a physics teacher. She stopped working outside the home for a while to raise the children, then became a magistrate. The mind-shaping moment of Rick's childhood came when he was ten years old and his parents gave him a Radio Shack TRS-80 computer, one of the first PCs. His father taught him to write software code in the Basic and z80 programming languages. From that moment on, Rick was hooked on science and technology.

Both of his parents were trained as physicists, and Rick got his undergraduate degree and PhD in physics at the University of Birmingham. During his studies, he began to understand the importance of systems in determining how the world works and how to make sense of it. In physics, self-organization takes place when some form of order arises from interactions between components of a disordered system. His next stop along the path to a full appreciation of the power of systems theory came when, at IBM, he began to use object-oriented programming (OOP) languages to build complex distributed computing systems—the kind that run ecommerce websites. The programming model revolved around objects, which contain both data and the software code providing instructions on what to do with it. The objects are designed to interact with one another based on the instructions in the code.

It was complex stuff, and Rick and his colleagues used an organizing tool called Design Patterns to capture the information about OOP designs that was most useful. To Rick at the time, it was surprising that the concept of Design Patterns had been developed by Christopher Alexander, a famous architect and urban designer. Alexander argued that the process of designing a town or even a major urban development project was too complex for one person to get her head around. It had to be broken down into discrete components, or systems, each to be designed by a different sort of planner or designer, and then the components would be rolled into a master plan.

Alexander was a bit of a polymath. His ideas also influenced the development of the first wiki, used for sharing Design Patterns, and the agile software development discipline—a technique that, among other things, enables teams of programmers to spot problems early in the design process and fix them on the spot. (More about agile in the next chapter.)

In the Smarter Planet world, Rick and his colleagues began to use systems thinking to help city leaders solve problems and to understand the complex relationships between transit systems, power grids, storm sewers, and the weather—for instance. You can see where this is going. Suddenly, systems were popping up everywhere. None of them operated in isolation. Understanding systems and their interactions was the key to understanding everything, from physics to computer science, to cities.

During those Smarter Planet days, Rick started writing a blog about what he was learning about urban systems, The Urban Technologist (https://theurbantechnologist.com). It became quite popular—landing him an opportunity to make a presentation to a UN commission and an interview by Al Gore. It's also how he met Peter Head. They corresponded about one of Rick's blog posts and later traded speaker gigs at conferences they were involved with.

The Smarter Planet campaign burnished IBM's brand and contributed significantly to public awareness of the value of sensors and big data analytics. IBM research produced a number of significant innovations—mostly in the areas of data integration and analytics. But Smarter Planet never became a big business for IBM. Initially, cities were strapped for cash, and stimulus funds went to traditional infrastructure projects. Later, the initiative was overtaken by the rise of cloud computing and artificial intelligence (AI).

Rick left IBM in 2014 and went to work for a succession of urban planning and infrastructure services firms in the role of smarter cities leader—developing new projects and executing on them. It was his dream job come true. While Rick was satisfied with this work, he found himself increasingly concerned about the direction of society and the economy. During most of his life, he had not identified strongly with the environmentalist label, although he had cared about environmental issues. Politically, he had been a moderate. In recent years, however, his politics shifted left. He began looking for bolder solutions to society's problems. He rankled at the underfunding of public schools in Britain and at economic inequities. He came to believe passionately that simplistic capitalist assumptions about economic growth driven by profitable enterprises that would deliver the outcomes we need as a society were mistaken. He worried that the country was not prepared for the economic dislocations he saw coming with the rapid adoption of AI and the resulting job losses. On the environmental front, he worried about the world that his twelve-year-old son, Tom, was inheriting. When he read Colin's email, he was ready to do something about it. "What more important reason could there be to work with this group of people and try to make a difference?" he said.

Andre's Journey

Andre, the fourth member of the original core team for Pivot Projects, had grown up in the shadow of a famous father—but he didn't seem to have suffered much as a result. Like Peter, he projected calm and unflappability. Also like his father, he was intensely curious about how things work. As a kid, he would pick up broken electronics devices at rummage sales, bring them home, take them apart, and try to fix them. Having a father who is an engineer gave him confidence that he could fix anything. He recalled that his dad would work out engineering problems on the family's dining room table by building models. Andre would arrive home from school to find Peter fashioning a tiny bridge out of plastic drink straws and rubber bands.

In university, Andre had set off in a different direction than his father. He received degrees in product and industrial design, but he didn't work in the field. Instead, in the late 1990s, he pursued website design. He lived in a big house full of young people in London and made a living as a freelance web designer. His first project was for a guy who ran a tiny music label. Later, working for a web design company, he helped create websites for big brands, including Apple. Then he worked for ad agencies for several years. When his dad left Arup to start Ecological Sequestration Trust, he volunteered for the organization in his free time, and later he managed the organization's website, analytics capabilities, social media, and collaboration platforms. "I was looking for a job that had more meaning," Andre said. "I was fed up with sitting in meetings with a lot of people with the word 'director' in their titles arguing about trivial things."

Ecological Sequestration Trust and Resilience Brokers were made up of the same small group of people who put on different hats depending on who they were meeting with. It was a virtual organization. Everybody worked at home. One of Andre's jobs was shopping around for a suite of communications and collaboration tools that suited the organization. Over time, he refined the portfolio, ending with Slack for real-time communications, Trello for organizing and sharing information, Google Docs for managing documents, and Zoom for videoconference calling. When Pivot Projects came into being, he got the group up and running on the collaboration technologies in minutes.

Bringing the Team Together

Putting together the collaboration platform was the easy part. The core team had a daunting task. They had to manage a group of people scattered around the globe performing an amorphous, complex, multistep task that would take months to complete. (Peter said that they were really facilitators, not managers.) These were volunteers. They could quit at any time. For some, English was their second language. Meanwhile, the leaders also had to raise money, reach out to potential partners, and in many cases perform day jobs at a high level. And, oh yes, this was happening during a global pandemic and a global economic crisis. They couldn't travel. They couldn't meet people face-to-face. They had to avoid getting sick. They weren't getting paid for this. And to top it all off, none of them had ever managed anything like this before.

Colin, Peter, and Rick divided responsibilities and went at it. Actually, they never really had a discussion of who would do what. They knew each other's skills and interests. It seemed obvious. Peter would handle high-level vision, relationships with policymakers, and raising money. Colin would be the day-to-day operations manager. Rick would set up some of the processes for getting things done and, as one of the workstream leads, would model how the subgroups should operate.

They shared a management philosophy: give people broad directions about what you want them to accomplish and suggestions about how they might accomplish it. Then grant them free rein to find their own way, as individuals and small groups. Check in frequently but in a mild-mannered way to see how they're doing. Make course corrections using guidance rather than commands. If something goes wrong, help them. Don't blame them. Inspire them. Give them hope. This approach, of course, was antithetical to the way organizations and initiatives are typically run.

Fortunately, the leaders had a bit of wind in their sails. As Pivot Projects took shape, there seemed to be strengthening support in the world for taking action on climate change. The United Kingdom had committed to reducing greenhouse gas emissions by 50 percent from 1990 levels by 2025 and by 100 percent by 2050. In May 2019, the British Parliament had declared a "climate change emergency." Meanwhile, in the United States, an April 2020 survey conducted by researchers at Yale University and George Mason University showed that a record-tying 73 percent of Americans thought global warming was happening and 66 percent felt a personal sense of responsibility to help reduce it.[4]

True, the Trump administration and other autocratic, antiscience regimes around the world seemed to be determined to destroy the planet as quickly as possible. But there were hints that the political will was changing. In summer 2020, with COVID-19 still spreading rapidly in the United States and the economy on the ropes, Joe Biden, the Democratic Party's choice for president, had seemingly taken on the mantle of environmentalism. He announced a $2 trillion climate plan that he expected to create millions of high-paying jobs, help the country recover from COVID-19, and help address systemic racism.[5] Pivot Projects leaders hoped that if Biden won and the Democrats swept the congressional elections, it was possible that the United States would be primed to do more to protect the planet than at any time since the birth of the environmental movement in the 1970s. The most powerful country in the world could lead a march to forestall the worst impacts of climate change.

Did the leaders of Pivot Projects have the skills, vision, and structure they needed to take the opportunity that lay before them and run with it? Could this fly-by-the-seat-of-the-pants, make-it-up-as-we-go-along group help slow the advance of climate change when armies of climate scientists and legions of UN bureaucrats had not? Was luck on their side?

PROFILE

Shulamit Morris-Evans

British Extinction Rebellion activist

On December 28, 2018, Shulamit Morris-Evans, then a twenty-three-year-old schoolteacher, was arrested along with two other women after they glued themselves to turnstiles and blocked the entrance of a London building where the directors of a coal company were meeting. The company had announced plans to establish a massive coal mine in Bangladesh that had the potential of displacing tens of thousands of people from their homes. The women were stuck to the turnstiles for four hours.

The arrest marked the completion of Shulamit's transformation from a mild-mannered student of the classics at the University of Cambridge to a person determined to help stop the destruction of the planet. Awakened to the damage being done by fossil fuels, she had joined a group called Extinction Rebellion, an international movement that uses nonviolent civil disobedience to prompt action to address climate change. "I went from thinking climate change was a solvable issue to realizing we're causing irreversible damage and long-term suffering and the dimming of human life on the planet," she recalled later.

Shulamit was one of the early joiners of Pivot Projects. Jeff Newman, an emeritus rabbi at her family's synagogue, was a friend of Peter Head. Peter had urged Jeff to join the project, and Jeff had invited Shulamit in turn. Jeff was also involved in Extinction Rebellion and had supported Shulamit during her court case. At the time, Shulamit was between things. She had left her teaching position and had traveled to Spain to work as an au pair and to learn Spanish. But with the spread of COVID-19, she had returned to London and was living with her parents in a spare bedroom.

In early Zoom meetings of the group, Shulamit was a near-ubiquitous presence. Behind her you could see clothing hanging on a standing rack—a symbol of the impermanence of her life. She joined the Faiths group. (She was considering becoming a rabbi.) Then she raised her hand to help Andre Head welcome, onboard, and train new volunteers. "I want to facilitate cross-pollination," she said at the time. Quickly, though, the group's leaders invited her to join the core team, treating her as an equal. She joined many workstream meetings—including education, sustainable infrastructure, and economics. She felt she had a lot to learn there, but she also probed and pushed the others.

Discussions in the economics group both excited and frustrated her. "I keep beating my fists against the economic structure in which we're operating," she said. "We need to go in one direction, but the structure is barreling us towards a cliff. It frightens me."

A few weeks into the project, during one of the weekly all-hands meetings, somebody expressed the desire to have less-formal ways for the participants to engage with one another—water-cooler-style conversations. So Shulamit organized a weekly spontaneous chat. Actually, she did two each Wednesday, one in the early afternoon London time and the other in the evening so she could include people from around the world. They were a hit. Sometimes six to eight people would join the Zoom chats.

Conversations veered all over the place, from the trivial to the profound. During the second chat of the day on June 24, 2020, for instance, a small group talked for nearly two hours. They touched on an amazingly wide range of topics, including the Black Lives Matter (BLM) movement in the United States, commercial real estate, the fact that many soldiers in Wellington's army at Waterloo were Irish, social media storytelling, the difficulty in overcoming bad habits, telehealth, and some of their kids. "I feel that one of the most powerful things to come out of this so far is the community it has created," Shulamit said later. "It's a community unlike anything I have ever been part of."

In the early stages of the project, Shulamit's mood tracked the instability, confusion, and hopefulness of the group as a whole. Many of the volunteers weren't sure what would happen next or even what they wanted to happen next. Still, she felt optimistic. "I hope it accomplishes what it's setting out to do," she said. "I hope we can help produce a change of global direction that is decisive, well evidenced, well thought out, and leads to the world we're hoping for."

3

The Scrum

Man *lives* on nature—means that nature is his body, with which he must remain in continuous interchange if he is not to die. That man's physical and spiritual life is linked to nature means simply that nature is linked to itself, for man is a part of nature.

—KARL MARX

If you had been dropped by a time machine into the middle of the Pivot Projects Zoom call on April 13, 2020, you might have wondered what was going on. It was a diverse crowd in some ways—perhaps twenty people in all. Among them was Awraham Soetendorp, a reconciliation advocate in the Netherlands, who had been one of the "Hidden Children" sheltered from the Nazis in Christian homes during World War II. Sarah Joseph, wearing a hijab, is an English writer and broadcaster who had converted from Catholicism to Islam at age sixteen. Frank Dabba Smith, an American rabbi living in London, is an advocate for peace in the Middle East and a historian of photography. Mama D Ujuaje, originally from Africa, is a chef and teacher who described herself elsewhere as a "plant whisperer." Martin Palmer, an investment counselor to religious organizations, had once advised Pope Francis on economics. The participants on the Zoom call touched on a wide variety of topics, ranging from the COVID-19 pandemic and its impact on people around them to systems thinking, to Doughnut Economics, an approach to making economies more inclusive and for measuring the sustainability of economic systems.

What connected everybody on the Zoom call was the fact that they were people of faith and that they were interested in Pivot Projects. This was the first meeting of the Faiths workstream, and it was typical of the first meetings of a number of the project's workstreams: there were a lot of introductions and exploratory conversations. Also typical was the fact that participants quickly elevated their observations to the level of the heartfelt and the profound.

Awraham, for instance, had been involved in a successful campaign in the Netherlands to pressure the government to set aside 500 million Euros to support refugees in the Middle East and Africa. Its aims dovetailed with those of Pivot Projects—to help set humanity on a more humane path. "We are making plans to try to change the interface of the Earth. I hope we will come together and fructify. I believe the world is ready for it," he said. He urged people to keep their minds open. "By listening to each other, we're listening to God," he said. "God tells us that we may not be able to reach our goal but that doesn't absolve us of the responsibility to do everything we can to try to reach it. Let us not be dismayed. Listen to the still voice. It will enter into us and give us strength."

In the Torah, it is written in the Book of Genesis that God created the world in six days, but that creation was not finished. God left that for humans "to do," *la'asot* in Hebrew. They were responsible for completing creation, a task that would never end, and, in later Judaic writings, for repairing the world, *tikkun olam*.

Pivot Projects had been launched in an hour of extreme pain and uncertainty. All around the world, people were dying horrible deaths, many of them separated from their loved ones. Bodies were stacking up in cold storage—and sometimes out in the open. Jobs and small businesses were lost. Children could not go outside and play or go to school. People were depressed and anxious. In many ways, life as we knew it had come to a halt. Yet this global collaboration of volunteers had to find a way forward—in spite of the fact that some of its members were also depressed, anxious, bereft, and insecure about their livelihoods. Several were unemployed and hurting for money. One volunteer, Tu Anh Ha, a schoolteacher from Vietnam, had been stuck in Europe since COVID-19 had struck and was unable to return home. She was couch surfing with friends. At one point, she was in the Netherlands, where she knew no one. Someone she met gave her a bicycle so she could get around.

In August 2020, a hurricane struck the east coast of the United States. Power was knocked out for hundreds of thousands of people—including some Pivot Projects volunteers. Many libraries and coffee shops were closed due to COVID-19, which made it tough to find a place with Wi-Fi, and communications were fitful for days.

The world was in need of repair and renewal.

There were tensions within the group as well. The sciences, the humanities, and the faiths jostled against each other from time to time—in good-natured and constructive ways. Sometimes, it seemed like they spoke in different languages. But even more fundamentally, there were gaps between those in the group who wanted to focus on producing practical solutions that would put the pre-COVID-19 world on a more just and sustainable path and those who advocated a much more radical approach: let the world collapse and pivot to something better. These conflicts typically operated below the surface, but they were there.

Stephen Hinton was one of the radicals. An Englishman living in Sweden, he had a career in corporate management rolling out telecommunications infrastructure, managing supply chains, and selling large-scale water purification systems. After he retired, he served several nonprofits addressing sustainability, including the Swedish Sustainable Economy Foundation. Stephen identified as a Marxist, but not the academic kind. He believed that capitalism is a deeply flawed system that is on the verge of collapse because the singular focus on maximizing profits and consumption is causing immense suffering and leaving behind vast swaths of humanity—not to mention destroying the natural world. He pointed to the fact that there are plenty of resources on the planet to make sure that nobody goes to bed hungry at night, yet more than 1 billion people do so. "Enough. We have to stand up and say, 'I know what it means to be righteous,' " he said. "Unless we connect with that, we're screwed. We're heading for extinction. We have this gift of intelligence and compassion. We need to steward ourselves, each other, and the planet."

Structure, Processes, and Tools

Rick Robinson, one of the three founders of Pivot Projects, attempted to give it structure and momentum from the start. He had worked in a number of performance-oriented organizations and was taken with the agile project management discipline, which had emerged from the software development world but was by then being practiced by all sorts of organizations. Central to the agile philosophy are the ability of participants in a project to collaborate with one another without hierarchies or rigid rules, to be extremely flexible and adaptable in their thinking, and to share a keen sense of purpose and urgency. In some forms of agile, groups start each day with a scrum in which each individual tells the group briefly what she or he is working on, what they hope to accomplish that day, and what could block them. The term *scrum* is borrowed from rugby football. A scrum is the phase of a rugby match were players from two teams pack closely together with their heads down and push against one another in an attempt to gain possession of the ball. It's the beginning of a session of play. The scrum was an inspiration for Rick's thinking.

Rick recommended that each workstream meet weekly and have both short- and longer-term objectives. They would report on their progress weekly on an all-hands Zoom call, where they would also get updates from the leaders. The workstreams would have a clear set of marching orders but also broad latitude in how they chose to get things done.

As things evolved, the plan and process for getting work done began to look like this:

- Get to know one another and decide on roles and meeting times.
- Start making conceptual models of significant systems within your domain, with a focus on issues of sustainability and resilience. Choose 100 core concepts and begin to put them into spreadsheets linked to the Kumu visualization tool.
- Choose a handful of key questions that it are critical for your group to answer—and that promise to lead to improving sustainability and resilience.

- Keep refining the questions and the models as you consider evidence and debate priorities, and use the Kumu visualizations to understand your domain more deeply and to share knowledge with one another.
- Load the models into SparkBeyond's artificial intelligence (AI) platform.
- Engage interactively with the AI platform to explore answers to your questions, and thus identify novel solutions to problems.
- Validate your conclusions with experts in the fields within your domain.
- Share findings with policymakers and communities.

I will explain the Kumu tool in detail in chapter 5 and the AI platform in chapter 8, but for now, here's a little about the technology underpinnings of Pivot Projects.

First, consider the status of innovation in the early twenty-first century. Typically, we rely on small teams of individuals to produce advances in science and technology, to develop revolutionary new products, and to create new and better ways of running businesses and delivering public services. Increasingly, innovations are produced at the intersections between multiple domains—such as biology, chemistry, and high-performance computers, in the case of medical science. Thus, the teams that work on innovation projects often are made up of people with different knowledge and types of expertise. Often—and more often since COVID-19—members of a team are not in the same room, building, or city. To do their work well, they need to share knowledge and data and to communicate with one another seamlessly using digital tools.

This is where twenty-first-century information technologies come in. Many of the software applications, data, and communications tools that teams of people use to get their jobs done reside in vast cloud computing data centers operated by companies including Amazon, Google, and Microsoft. It is a highly efficient way of getting computing done, but even more important for innovators, it makes it easier for them to share information and to collaborate.

Within the cloud universe is a cornucopia of services that focus on helping people accomplish particular tasks. Andre Head, the technology leader for Pivot Projects, chose Zoom for videoconference calls; Slack for communications and collaboration; Trello for organizing and sharing how-to videos, software tools, and documents; and Google Docs for storing documents and data. In addition to those communications and sharing tools, the group chose two tools designed to help people think: Kumu and SparkBeyond.

Kumu (www.kumu.io) is a visual language that helps people organize their thoughts. Using Kumu, individuals and teams don't have to stare at endless rows and columns of data in spreadsheets or use query languages to search databases. These latter methods make it a struggle to identify patterns and make connections. Instead, Kumu embellishes existing spreadsheets or creates new ones and transforms tabular data into visual information. This is done using the Create, Select, and Drag and Drop functions we all have become accustomed to in personal computing. A typical Kumu diagram shows an array of color-coded dots, called nodes, usually with text labels attached to them, which are connected to one another by lines, called edges.

For Pivot Projects, the Kumu nodes were divided into three categories: resources (or topics), processes, and attributes. It might be helpful to think of them as nouns, verbs, and adjectives or adverbs. By converting the models they were building into visual forms, the Pivot Projects teams created knowledge maps. By manipulating the nodes and edges on the maps, they could better understand the relationships between the nodes, including causal relationships.

To make this approach to organizing information work, Pivot Projects needed a very powerful AI platform. Fortunately, Peter Head was introduced to an Israeli AI company called SparkBeyond by Nigel Topping, the United Kingdom's High Level Champion for Climate Action for the United Nations Climate Change Conference of Parties (COP26). SparkBeyond's AI machine uses machine learning and other techniques to help humans develop hypotheses and solve problems. People at SparkBeyond had conveniently designed the AI system to ingest information from Kumu spreadsheets. As a result, the Pivot Projects teams would be able to develop systems models with Kumu and later load them into the SparkBeyond machine, which, like all the other technologies described here, operates in the cloud. The machine had already been loaded with a vast amount of other information, including everything published in the Wikipedia database, patent listings, academic and scientific reports, government databases, and articles published on the internet.

By loading the Pivot Projects' models into the SparkBeyond machine, the machine would be equipped to enter into what was essentially a digital conversation with people from the Pivot Projects. People could ask the machine questions, view a variety of answers, choose the most relevant ones, explore underlying data, ask more questions, and so on. The answers to the questions would typically be presented in a visual form, called a knowledge graph, which, like Kumu diagrams, helps people see and understand things and the relationships between them. In this way, they would be able to focus on interesting ideas and innovations that might be useful for making particular cities or regions more sustainable and/or resilient.

When the Pivot Projects participants began to meet, the interactions with the SparkBeyond machine were months in the future. Still, it was important for them to think ahead about how they would eventually interact with a sophisticated AI machine.

Getting Started

At the first all-hands meeting, which was a Zoom call that took place on April 3, 2020, the meeting crackled with excitement. Fifty-five people were on the line, most of them friends and associates of Peter Head, Colin Harrison, or Rick Robinson. It was clear that these accomplished people, who had been busy solving big problems within their own areas of expertise, enjoyed pulling together with other like-minded people to take on some of the biggest problems. Colin explained to them how the whole thing started with a bit of self-deprecating humor: "I was sitting around on a Saturday afternoon thinking about how we might save the planet—the way one does."

The volunteers took turns introducing themselves and raised some of the issues that they thought were most interesting or confounding.

Group Zoom call on April 3, 2020

They talked about everything, from pandemics, open data, global supply chains, and grassroots engagement to feedback mechanisms for behavioral change and having the courage to ask "dumb questions."

They were harshly critical of the status quo in the world and did not want to return to it after the danger of COVID-19 passed. Robert Bishop, president of the ICES Foundation in Switzerland, which funds the building of models of the world's natural systems, said, "We built a global economy that was wound up as tight as a drum. It only took what was really a minor perturbation to split it into disaster. That's what the virus has done. That's the unnaturalness of what we have done. As we move into the future, the economy has to be married into nature. It cannot be as wound up as it was, with no reserve."

They explored the challenges of making sense of complex situations. Christopher Barrett, a biocomplexity expert on the faculty of the

University of Virginia, warned of the fallibility of system models, no matter how carefully they're created. He said, "For every class of decision in a given circumstance, the very notion of what's relevant and what you can rely on changes. One of the things that has made science so difficult to apply to the world in environments that involve social and technological factors is that people are looking for a book of facts. Even scientists sometimes are looking for a book of facts. But, in fact, what we have is a provisional-knowledge world. To deal with that, you have to follow an abductive process of guess, act, and test."

Mike Goodfellow-Smith, a research fellow in sustainable cities at the University of Birmingham in the United Kingdom, spoke about the importance of choosing the most persuasive language when making a case for a change in policy or practice. In the wake of COVID-19, he said, *mutual aid* is probably the best term to use when discussing investments in recovery because it fosters compassion. Citizens are willing to contribute or sacrifice to help somebody else if they feel empathetic toward them. "We want to stimulate the right kinds of investments," he said. "We might be more successful if it's framed as mutual aid and taxpayer money for taxpayer benefit, not for corporates."

That first all-hands Zoom call was just a taste of things to come. The early meetings were often wonderful, free-flowing discussions. In some ways, they were like cocktail parties at the faculty club of a university—a bunch of smart people were thrown into a virtual room together and they had a great time getting to know one another.

They were also touchingly personal at times. Participants talked about their children and grandchildren, and about their love for the planet and humanity. Alan Dean, who headed the Education workstream, once commented, "My daughter asks me why I get involved in all of this enviro stuff," he said. "I tell her, 'I'm not leaving you the house, but I want to leave you something more important: hope.' " James Green, a recent graduate of a master's degree program in environmental policy who was working as a record store clerk, said that since the COVID-19 pandemic started, people who came into the shop were behaving differently. "People unburden themselves to me like I'm a bartender. They're open about their feelings," he said. "Middle-aged men talk about how much they enjoy being home with their families. They're less interested in work. They're reevaluating their lives and their values."

At other times, the volunteers struggled to get their heads around the enormity of their task. Would they actually be able to create a whole-Earth model? How do you choose just 100 words to begin to represent an entire domain of knowledge in all of its complexity? That was particularly challenging for the Faiths group. An initial round of brainstorming produced more than 500 key words. On further reflection, they decided they were missing some and swelled their list to more than 1,000. Then they became brutally reductive and shrank the list back to 75. Then they expanded it back to about 125. Most of the groups came face-to-face with one of the main challenges of collaboration in the digital age: the fact that the work of the hive brain can be messy and exhausting.

Those weekly all-hands Zoom meetings took place on Fridays. Typically, in the early days, there were about thirty to forty people on a call. They began with progress reports from the leaders and the heads of the workstreams and, after a time, they included group exercises aimed at raising awareness or developing a consensus around group goals and priorities.

The volunteer staffing situation at Pivot Projects was like the weather in the New England region of the United States: always changing. Some of the high-profile friends of the founders dropped out after a couple of weeks—too busy or not interested. Meanwhile, there was a small but steady stream of new volunteers. People invited their friends, their work and professional colleagues, and sometimes random strangers.

Richard MacCowan, a Scot and leader of the Ecology and Planetary Health workstream, scouted for new members when he attended webinars on related topics sponsored by other organizations. He would write down the contact information for others who were on the calls and email them later with an invitation to join Pivot Projects. In this way, he recruited nearly a dozen people to his workstream—many of them from India.

Richard was one of perhaps two dozen people in the early days who had an abiding passion for the project and put an immense amount of time into it. These people joined multiple workstreams and could be counted on to show up for the Friday all-hands meetings. In some cases, and Richard was one, people did professional consulting work that was closely related to the project, and their involvement helped them make connections that might land them commercial jobs in the future. In many cases, however, there really wasn't a commercial payoff: their motivations were purely altruistic.

There were many unconventional thinkers in the group. Perhaps they self-selected because they were willing to go on a wild, uncertain ride with a bunch of strangers in an attempt to save the world. For instance, Bill McKenna, an American from Lubbock, Texas, currently living in Austin, said *this* when the group was talking on Zoom about how to reach leaders in cities who might want to engage with Pivot Projects: "We could reach out to the addiction recovery community. All kinds of people are there; leaders, too. They are in recovery and they are pivoting, personally. Our project might resonate with them."

The Zoom meetings themselves were often great fun. We met all sorts of people and they came from all over. One participant, Ian Mabbett, a chemistry professor in the United Kingdom, sometimes dialed in from a recreational vehicle in his driveway. Another participant, Manesh Sah, a student in Nepal, called in using a borrowed smartphone. You could see the Himalayas behind him. Richard appeared in a tiny, well-lit room where he was surrounded by twenty-seven house plants. He once demonstrated to a small group how you can drop a banana peel into a jar of water and two days later water your plants with the nutritious liquid.

From the start, though, the pool of volunteers skewed older, white, male, and Northern European and North American. That was a problem. How can you develop practical solutions to address the world's problems if many types of people and places in the world were underrepresented or not represented at all? Efforts to bring in young people were successful to some extent. The project attracted young folks from India, Dominican Republic, Bangladesh, Indonesia, and some countries in Africa, in addition to Northern Europeans and Americans. However, while a few of the young people were outspoken and engaged, others, especially those from developing nations, tended to hang back. They faced language barriers and Internet connectivity challenges. As the project continued, several people in the group vowed to take it on themselves to reach out to people who could add diverse perspectives.

Digging In

The twenty-plus workstreams were the soul of Pivot Projects. The leaders chose the workstreams by identifying topic areas that were relevant to COVID-19, sustainability, and resilience and then combined them or carved them up in ways that produced, in the end, an incomplete jigsaw puzzle of interlocking pieces. These groupings of people and concepts would produce the models of human-built systems interacting with natural systems, which would connect into the whole-Earth model.

A number of the workstreams met regularly and were generally well attended. There might be six to eight people on a Zoom call. Others had trouble getting off the ground or getting much done. Individuals would join a group, listen to a couple of sessions, then disappear. People would volunteer to lead a group and then discover that they didn't have the time for the assignment. It fell to Colin to try to keep all of the workstreams engaged and moving forward at a deliberate pace—which proved to be a nearly impossible task. He was like a plate-spinning acrobat in an old-time circus. Somebody once quipped that Colin's job was like "herding invisible cats."

The leaders had originally hoped that a notable expert in each field would lead each workstream, and that came to pass in a number of cases. For instance, Peter Williams, an expert in water management systems, ran the Water, Floods and Waste workstream, and Stephen Hinton headed the Sustainable Infrastructure workstream. But having recognized domain experts lead the study groups became the exception rather than the rule. What we often had instead was people with a broad array of interests taking on leadership roles for groups that were related in some way to their areas of expertise. For example, Damian Costello, an Irish business consultant who specialized in disruptive innovation, lead the Economics, Law and Politics study group.

Colin Harrison

There was a bit of unease among some of the leaders with this state of affairs at first, but they quickly warmed to the path the project was taking. The heads of the study groups tended to be synthesizers. They viewed the world holistically. In human resources parlance, they were T-shaped people, with a wide variety of hard and soft skills, as opposed to I-shaped people, who tend to go deep in a single area. The T-shape metaphor is also used within the agile project management discipline, where it's considered to be important to mix people with a variety of skills and domain knowledge in a scrum. Another thing that set the managers of the workstreams apart was that they tended to be highly motivated to challenge the status quo and conventional thinking—something that you might not find in somebody who sat comfortably within a domain. They looked at the world through radical eyes.

The Pivot Projects leaders also figured that there were opportunities to access domain expertise in other ways. The groups would tap into mounds of research studies across a wide array of fields as they built their system models. Then, when the groups began to interact with the AI machine, a lot of domain knowledge would be programmed into the computer. In addition, the groups would consult with outside experts when they engaged with people in cities and regions. "This is the essence of collaborative intelligence. It can accomplish things way above what any individual can achieve," said Peter Head.

The workstream meetings tended to be free-form and unfocused at the start. Despite the weekly cadence and structure that the leaders had set up to encourage productivity, a number of them typically started off with about twenty minutes of lively but often off-topic discussion before the participants focused on the job at hand in the remaining forty minutes.

A handful of the workstreams were highly organized and results driven from the start. In some cases, this was due to a heavy influence from what I'm calling the IBM Mafia. About one dozen former or current IBMers were involved in the project in one way or another. Colin was the most high-profile of them in the smart-cities domain, but they included a couple of other former executives, including Gerard Mooney, a key player in corporate strategy at IBM. Jim Cortada, a former IBM consultant, had authored a number of books about the history of computing and information—so he had deep expertise in critical areas for the group. Several of the key thought leaders from the Smarter Planet campaign were also involved, including Peter Williams, Ian Abbott-Donnelly, Stan Curtis, and Mark Cleverly. A number of the IBMers had worked together on strategies or Smarter Planet projects, so they knew each other's skills and personalities, and they knew how to get stuff done.

Next, I'm going to explore four of the workstreams in some detail so you can see how they functioned and how they related to one another. I'm starting with one that's relatively narrowly focused, and then I will progress with each example to ever more expansive systems.

Water, Floods, and Waste

The Water, Floods and Waste workstream was one of the overachievers. Peter Williams ran it and got a lot of help from Ian Abbott-Donnelly and another IBMer, Jean-François Barsoum. They had worked together for years on IBM water management projects. Several other people who had no IBM connections also made significant contributions to the ideas that the Water group developed. The group quickly collected the first 100 words for their systems model. Ian, who loved exploring new technologies, took it upon himself to master Kumu. He used the Water group as his test case.

The group quickly charted the scope of its deliberations and the issues it would explore. It would focus on three areas: the dynamics of supply and demand for water, the socioeconomics and sociopolitics of water, and what it would take to reengineer water and wastewater systems. They quickly began to spot patterns that could have a dramatic impact on how water is provided, paid for, and treated.

One such pattern was tied directly to COVID-19. During the COVID-19 pandemic, commuting to work came to a screeching halt. In the hardest-hit countries, roads, trains, ferries, and subways were just about empty. It was expected that much of the moving around would resume after the danger of COVID-19 passed, but what if it didn't? What if large service businesses based in cities decided it was okay for many of their employees to work from home on a long-term basis. That would change settlement patterns, and thus water-use patterns. Today's water and waste systems would have excess capacity in some places and shortages in others.

The group concluded that city planners should begin thinking about adapting their water systems to deal with the emerging realities of the post-COVID-19 world. Today's water systems are centralized and based on capturing economies of scale. They have large pumps, pipes, and treatment facilities. They require huge amounts of energy to move water around and they're vulnerable to disasters. If a giant sewage treatment plant is flooded by a storm, many thousands of people and large bodies of water are affected.

Instead of pouring more money into this expensive and vulnerable way of doing things, why not break water and wastewater systems down into smaller units? They'd be like microgrids in the electrical utility world. Smaller-scale systems would be more energy efficient and less vulnerable. They could be designed and built quickly to meet sudden changes in living patterns.

The group broke down its explorations into ten subcategories, explained the post- COVID-19 dynamics for each, and prepared a list of questions they wanted to investigate in collaboration with the SparkBeyond AI machine. "We hope the AI will help us make connections and identify solutions that we weren't aware of," said Peter Williams. "I don't expect to bump into something that's dramatically new, but I could always be wrong. It would be huge fun to be wrong. I'd love it. Those could be game changers."

Sustainable Infrastructure

Adjacent to and overlapping with the Water workstream was Sustainable Infrastructure. The group surveyed infrastructure for transport, energy, housing, food provisioning, water, healthcare, waste handling, and money payments.

Stephen Hinton ran this group with an easygoing yet energetic style. He was constantly throwing out provocative ideas, often connected to a book he had read or a group of save-the-planet people he was involved with. He was funny in a self-deprecating way. Not your typical Marxist, probably. He started off the discussion during the group's June 4 Zoom meeting by giving his colleagues an extremely challenging factoid to chew on: the 2018 Report of the Global Commission on the Economy and Climate had estimated that the global price tag for infrastructure investments to meet the UN's Sustainable Development Goals (SDGs) was a whopping $90 trillion.[1] Where would the money come from to finance this huge endeavor?

This discussion evolved into one of the group's four key hypotheses: the idea that too much of the world's capital is being invested in purchasing existing infrastructure rather than in creating new and more energy-efficient and sustainable infrastructure. Capital was being spent on buying assets rather than developing new capabilities. Not enough of it was going into initiatives that reduce energy use or create jobs. If this capital was redirected into more productive investments, there would be plenty of money to pay for the infrastructure needed to fulfill the SDGs, the group concluded. They wanted to prove the hypothesis by obtaining more data and studies and by engaging the SparkBeyond AI machine in an exploration of the possibilities.

Stephen already had some firm ideas of what had to be done on a macro level. He wanted to see national governments set up initiatives aimed at investing public money wisely in new infrastructure, and he wanted to see legislation and rules that would encourage financial services companies and corporations to invest in the future rather than doubling down on the past. "Wall Street needs to be put in a box like a naughty dog," Stephen said, "And stock corporations need to have a fiduciary responsibility to society rather than focusing solely on maximizing returns."

Economics, Law, and Politics

This group took a wild ride. It started off focusing on economics, then expanded to encompass law and politics. It was led by Damian Costello, a voluble Irishman who developed his thoughts largely by trying them on Zoom meetings via long, passionate rambles. On calls, he typically ducked his head a bit, so you mostly saw the top of it.

Damian was a business consultant focused on disruptive innovation. He advised company executives on how to avoid being overtaken by upstart companies with better ideas and how to generate game-changing innovations internally. He had grown disenchanted in recent years after he observed that many corporations seemed to be more interested in suppressing innovation than fostering it. Time after time, he saw established players in industries purchase the innovative upstarts that threatened them and then fail to capitalize on their innovations.

At the same time, he believed (as an amateur but avid student of economics) that capitalism was ripe for a fall. The neoliberal regime that had emerged after World War II with a focus on consumerism had a tight grip on the global economy, but the wealth created by the system was increasingly being accumulated in the hands of a few, and the forces of production it had unleashed were laying waste to the environment.

Neither situation was sustainable, and he expected a major collapse at some point—and thought that the COVID-19 pandemic might set it off.

The group embraced his thinking as a taking-off point and, over time, it developed an elaborate model that contrasted the capitalism of the present, which it called toxic prosperity, with an alternative economic system for the future, which it called healthy prosperity. This new model for prosperity would value the well-being of people and the planet over the interests of capitalists and their allies in government. Its goals were to achieve fairer distribution of wealth and opportunities, and regeneration of damaged ecosystems.

The relentless drive for consumption of goods and resources was the villain in this drama. The consumerism that took off in the 1950s was partly about people desiring to live a good life as measured by material things and to give their children a better life than previous generations had experienced. That impulse was combined with the drive by companies and investors to make ever-larger profits. The result was overconsumption. Overconsumption was like an opioid. It felt so good at first, then became an addiction. It was destroying the planet and causing great suffering. "We're addicted to a prosperity that is slowly killing us," said Damian.

A new kind of consumerism was focused more on doing good than on acquiring goods, and on reciprocity more than dog-eat-dog competition for resources. The idea was that, by being mindful of the value of shared well-being and a shared planet, people might modify their behavior and live in ways that benefitted others and the environment as well as themselves. People who were well off would consume more responsibly and would use their consumption as a tool to reward economic actors who were in sync with their values and punish those who were not. On the supply side, corporations would recognize that only by radically changing their values and operations would they be able to survive over the long haul. Just like industrialist Henry Ford had seen that only by giving his workers a living wage would they be able to buy the cars his factories produced, today's industrial titans needed to see that only by treating people fairly and improving well-being would they be able to grow their revenues and profits over the long term. It all sounds pretty idealistic and perhaps impossible to achieve, but members of the group figured that there was no harm in planning a world that would be dramatically different and plenty of harm in settling for the status quo.

Damian saw the group's work within the framework of his innovation practice. Each major disruption of industry comes when a new innovation emerges that is superior in fundamental ways to the technologies, products or services, and business models that preceded it. Think of Apple's iTunes and iPods, which upended the music industry. He thought the same kind of disruption could happen in the global economy, which is, essentially, the world's uber industry.

This view of the future of economics came together after the group had gone through its first round of questioning and model building in Kumu. The next steps for the group would be adding these concepts and interdependencies to the model and adding more material from the domains of law and politics. Damian had also given himself a task. He would write a brief similar to the ones he typically authored for his corporate clients. It would be aimed at convincing business executives that they would be better off over the long run by shifting aggressively to healthy prosperity rather than being dragged kicking and screaming into a world that would be increasingly hostile to their aims and their methods.

Members of the workstream hadn't even begun to come to grips with a conundrum that faced everyone in the environmental movement: the fact that using gross domestic product (GDP) growth as the measure for economic success and well-being was incompatible with environmental sustainability and societal resilience.[2] Capitalism and its drive for economic growth was killing the planet and people. This was a sticky wicket, though. It was one thing to urge well-to-do people in developed nations to live more sustainably and forego GDP growth, but how to you tell that to a poor farmer working a half-hectare of land in India?

It was so clear that climate change and other natural disasters disproportionally affect the poor and downtrodden. This was a fairness issue. In the United States, the COVID-19 infection rates for disadvantaged Black and brown people far outstripped those for the population at large. At the same time, developing nations were feeling the brunt of the first wave of impacts from climate change. For instance, in the summer of

2020, one-quarter of Bangladesh was submerged by heavy rains combined with sea level rise.

How could the rich people and countries of the world work out a fair deal with the poorest Bangladeshis and their brethren around the world?

Faiths

In the earliest days of Pivot Projects, Peter Head had a particular job in mind for the Faiths group. He believed that when the organization began to engage with communities, the people in the Faiths group could reach out to faith leaders in the cities and regions and get them and their constituents involved. He imagined that local faith leaders could help convince people in their communities that a more sustainable future was worth sacrificing for and enlist them in taking the necessary steps to get there. In an earlier engagement by his group, Resilience Brokers, he had seen Christian faith leaders in Accra, Ghana, play that kind of role.

To coordinate the group, Peter had recruited a friend, Jeff Newman, a retired rabbi he had met when they worked together on the Earth Charter, a declaration of fundamental ethical principles for building a just, sustainable, and peaceful global society. After retiring from his post as the rabbi at a northwest London synagogue, Jeff had become an environmental activist and a campaigner for economic justice. He was inspired by prophetic Judaism, the stories in the Hebrew Bible of Isaiah, Jeremiah, and Amos, who taught the importance of justice and peace. Jeff was involved in a number of environmental protests—culminating in an Extinction Rebellion action in front of the Bank of England on October 14, 2019. He was arrested after he sat down in the middle of the street and refused to move. "We are in a period of enormous catastrophic breakdown, and, if it takes an arrest to try to find ways of helping to galvanize public opinion, then it is certainly worth being arrested," Jeff said at the time.

Jeff Newman's arrest in London on October 14, 2019

Jill Mead, *The Guardian*

Jeff helped organize the Faiths group and served as one of its leaders. He had a more expansive view of the role of the group than Peter Head had at first laid out. He came to believe that the group could help guide the entire enterprise—influencing the thinking of all the teams, reminding them of the importance of beliefs and ethics in shaping a new world and urging them to ask the big questions about human life. Asked what the Faiths group could contribute, Jeff said, "I think for many people it's easy to forget the numinous—the sacred, the sublime, the holy. We need to accept that we are not the center of the universe, that we are one small part of the totality of being. That's what faiths at best are all about. We need to be able to take that moment to consider why we are here."

The influence of the Faiths seemed to unleash expansive thinking among others in the group. While there wasn't much talk of God outside the Faiths meetings, people frequently spoke about the future of humanity and the meaning of our lives. Tom Rossiter, an architect and photographer, spent several days during July 2020 driving through the Rocky Mountains in the United States in a pickup truck loaded with sophisticated camera equipment, shooting the night sky. He lived just outside Chicago and so rarely saw the Milky Way. Now, it was his window into the infinite. On a mountaintop in Utah's Uinta Mountain Range, he was awestruck. "It was pure joy," Tom recalled later. "I had this sense of awe, wonder, respect, and the quiet of being." Awe was one of the words chosen by the Faiths group for its Kumu model.

The Faiths group sought to inspire their Pivot Projects colleagues by introducing them to people for whom spirituality, culture, and science were inextricably woven together—three indigenous grandmothers. They had been looped in by Paola Bay, an Italian artist and designer. On a Zoom call arranged by Paola, three grandmothers spoke about their religious and cultural traditions—which were all based on a profound respect for nature. Then they gave pep talks. "We're all in this together," said Susan Kaiulani Stanton, who was native Hawaiian and Mohawk. "Yes, it's chaotic out there, but chaos precedes change. Anyone who has ever labored on a table and given birth knows it is not easy and it's not fun, but a beautiful stretching of your heart takes place."

The Faiths group also brought thinking tools to help Pivot Projects members navigate the complexity and uncertainty of their tasks. This work was led by the co-leader of the workstream, Deborah Rundlett, the pastor in a Protestant church in Connecticut who also ran a change management consultancy. One of her focuses was on the Two Loop Theory, which was created by business consultants Meg Wheatley and Deborah Frieze for managing organizational change, but it had broader applications. The two argued that change in human systems often emerges from a series of disconnected local actions that link, gain strength and purpose, and go on to catalyze social change at a massive scale. Examples are the global peace movement and Extinction Rebellion, the climate change action movement.

The Two Loop Theory is expressed in a diagram that charts the transformation of an organization. The first loop traces the rise and decline

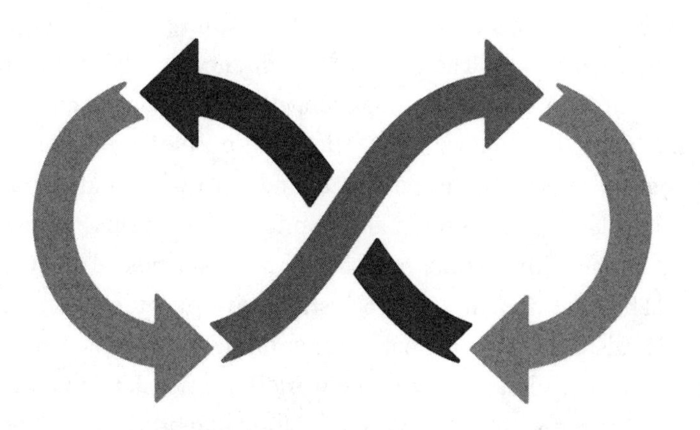

Pivot Projects logo

of the present state. The second loop traces the process of rebirth. A key element of the theory is that changemakers have to "hospice" the positive elements of the old system even as they are reimaging and creating the future system. Pivot Projects explored a number of theories of change, several of which represented reality in two-loop diagrams. Members of the group liked these models so much that they adapted the infinity symbol as their logo.

Debbie had managed quite a multidimensional career. She started on Madison Avenue on the advertising team that created the famous series of Charlie Chaplin ads for IBM's first personal computer in 1981. But she had felt a calling to the church since childhood, and after getting masters and doctor of divinity degrees, she served in pastoral or management positions in the Presbyterian and Congregational churches in Pennsylvania, California, Connecticut, Illinois, and Ohio. In many cases, she was in a turnaround role. An organization needed to pivot, and she helped manage the transformation. That led to her foray into change management counseling and to founding a group called Poets and Prophets, which is aimed at helping people lead social, environmental, and economic change. Debbie said, "I believe we're at the end of a 500-year cycle when it's not time for incremental change; it's time for deep adaptive change."

Members of Faiths also joined other workgroups, either as regular members or as visitors. The goal was to influence—so the scientists and social scientists in the other groups kept ethics, spirituality, and religion in mind during their deliberations. The Faiths people were acutely aware of the natural tensions between science and spirituality, and they sought to build bridges between them and also to explore them. For instance, Paola Bay, one of the members of the Faiths group, posted an article on Slack, "The Environment Is Not a System," in which the author argued that it's a mistake to view the natural environment in mechanistic terms. "When armed with a computer, the world becomes data and relations appear as systems. In this way, computational approaches like modelling and A.I. simultaneously reveal and obscure aspects of reality," the author wrote.[3] The posting gave rise to a spirited discussion on Slack. "I love to nudge and trigger people," Paola said. "We have a tendency to box and intellectualize nature. We think we know everything. We don't. Nature is more powerful than us."

The leaders of several other workstreams appreciated the contributions of the Faiths people. In the Water group, for instance, Jeff helped shape one of the main proposals—its view on water pricing. "Jeff and I really connected. We really hit it off," said Peter Williams, the group leader.

There's a conundrum in the water world: Everybody needs potable water, so it should be affordable to all, yet it's also a finite resource, which argues for pricing based on scarcity. Also, if something's free, people tend to waste it. Jeff had impressed on the group the importance of viewing water as a human right. In the midst of the COVID-19 pandemic, water was even more critical. Many poor people without access to clean water could not follow the protocol for handwashing. Peter said that Jeff helped the group design a framework for water pricing where a certain amount of water would be free for those who can't afford to pay and that the prices would rise on a curve based on how essential the use is to sustaining human life.

Three months in, Faiths members felt that they had contributed to the larger group in significant ways, but they also saw that much was left to be done. For starters, they needed to diversify. Most of the active members were from the United Kingdom or the United States. Most were Christian, Jewish, or Muslim. They agreed to reach out and recruit people from other religions, races, ethnic groups, and geographic regions. Second, they planned to organize themselves to begin engaging with communities, as Peter Head had first envisioned their role. Exeter, in the southwest of England, had expressed interest. Jeff had already planned to spend a weekend there with his wife. "I'll scope out the place," he said.

Breakthroughs

By midsummer, Pivot Projects was clicking. Most of the workstreams had finished creating their basic Kumu models, chosen the questions they wanted to ask the AI machine, and had prepared reports containing some of their early findings. Each of the Friday all-hands meetings seemed to produce a new shared insight that brought members of the group closer together and helped propel the whole enterprise forward.

One of these exercises produced the first clear-cut set of priorities for the project. The participants divided into small groups to discuss the question of the day: How would you prioritize outcomes from the Pivot Projects? The exercise forced them to think deeply not just about priorities but about the outcomes they wanted. What did they hope to accomplish and how did they hope to accomplish it? Here are the results:

The What

1. Create meaningful jobs and protect human rights by investing in green technologies and practices.
2. Stop polluting and start regenerating ecologies.
3. Improve air quality.
4. Provide safe water and sanitation for all.
5. Reduce greenhouse gas production.

The Means

1. Foster economic systems that focus on well-being, not on GDP growth.
2. Provide high-quality education for all.
3. Focus on local communities and listen to them.
4. Help regenerate local habitats.
5. Scale up innovations from local to global.

On another occasion, Debbie Rundlett from the Faiths group conducted a virtual retreat designed to be part pep rally and part brainstorming session. About forty people participated in a three-hour Zoom call. One part was presentations by a bunch of the workstreams. It was impressive how much progress had been made. They had put together detailed reports backed by abundant evidence. The community-building exercises were a lot of fun. For instance, during a storytelling session, Shulamit Morris-Evans, a member of the leadership group, wrote the lyrics of a song, "We Want to Change the World," in about five minutes flat. Then she sang them to the tune of "I Could Have Danced All Night" from the musical *My Fair Lady*. It was silly yet touching.

The retreat was also an opportunity for heartfelt reflection. Sarah Joseph, the writer who had converted from Christianity to Islam in her teen years, spoke about being daunted by the task that lay ahead for the group.

> Sometimes I get demoralized by the state of the world, but then I reflect that it was forever thus. This is the nature of the human condition. It's a battle between good and evil, and between greed and generosity. We can build social structures that support the best rather than the worst in us, but in the UK and the US today, decency isn't in control.
>
> My mother owned a model agency. I grew up in that world. I could see that my world was false even when I was tiny. My husband grew up in a mud hut in Bangladesh. The people in my husband's village want to come here. They think the streets are paved with gold. How do we convince people in the West that this dream is an illusion but also tell people in the Third World that it's an illusion?

Waël Alafandi, a young refugee from the Syrian civil war living in France, was a cheerful optimist whenever he spoke in meetings, and that day was no different. "Us coming together, having these experiences, the ideas—it matters," Waël said. "We can create a consciousness that can be passed on to people. We can plant the seeds, they will germinate, and they'll keep growing and growing."

While there was a lot of inspiring talk that day, arguably the most significant insight came from Jim Cortada, the former IBMer and book author. He reported for his workgroup, which was called Industries and Business but had veered off into another project entirely: developing a model for Pivot Projects engaging with communities. Most of the members of the group were ex-IBMers, and, like Jim, they had worked in business consulting—engaging with government and business leaders and advising them on how to make their organizations work better. The group had crafted an engagement model for bringing the best ideas generated by Pivot Projects to leaders of cities and regions, collaborating with them, and producing positive outcomes aligned with the SDGs.

Jim referred to a phrase former U.S. President George H. W. Bush had used several times before and during his presidency: a thousand points of light. Bush was talking about the power of volunteerism. He called upon this spirit to help move the country forward. During his inaugural address on January 20, 1989, he used the phrase twice and said, "The old ideas are new again because they are not old, they are timeless: duty, sacrifice, commitment, and a patriotism that finds its expression in taking part and pitching in."[4]

Jim told the group that if they were able to really connect with people in local communities, could get modest sustainability initiatives launched, and show some results, they could then offer them as models to be replicated in other places. A few successes could scale up to thousands. It could have real impact. "This is about converting ideas into practice," he said.

A couple of weeks later, a seemingly huge breakthrough came. Pivot Projects was invited to present its work to a video gathering of central government and local government officials hosted by the UK Cabinet Office. The group's focus was on using technology, including AI, to help deal with climate change. Peter Head would have five minutes to present, then there would be another twenty-five minutes for questions. He spent a week preparing and polishing the slide show and narration. "I've never had an opportunity quite like this before. We've never been able to speak directly to the government," he said a few days before the meeting.

The big day came, on August 7, 2020. Peter presented the Pivot Projects' story to about twenty-five government officials. Things went well, and they managed to ask five questions in the time allotted. Among them were: What kind of questions could the AI machine answer? What was the weakest link in the model? A woman from a local authority said every town needs the kind of capabilities that Pivot Projects was developing. A few minutes after the call ended, an official from the UK central government procurement office called Peter to say that they had just set up a system for connecting technology providers with towns looking for decision-making support. Pivot Projects could be put on it. During an all-hands meeting a few hours later, Peter reported back to the group. In his understated way, he showed a lot of optimism. He said, "I think we have a real chance to turn a corner in terms of moving ahead and having impact."

PROFILE

Waël Alafandi

Syrian refugee studying in France

One night in April 2015, Waël (pronounced Wah-el) Alafandi, then nineteen years old, climbed into a twenty-foot inflatable boat with fifty other refugees and made the journey from Turkey to the Greek island of Lesbos. The motor stalled in the middle of the trip and the boat was nearly swamped, but the group landed and they were taken into custody by Greek authorities.

So went one of the early passages of Waël 's six-week journey from his hometown of Aleppo, Syria, to Berlin, Germany. Waël and a handful of his friends traveled together. They caught a few rides on trains and buses, but they moved mostly on foot—walking along rail lines by night and sleeping in forests by day.

Five years later, in the spring of 2020, Waël's life had been transformed. He was married to a French woman and living in the south of France, where he participated in a year-abroad program as an extension of his studies at a university in Berlin. He was also a Pivot Projects volunteer.

He had found his way to the project through Alan Dean, the leader of the Education workstream. They had met at a peace conference and Alan had urged Waël to join. "I believe in making positive change," Waël said. "I see how important it is for people to come together. I'm humbled to be part of it."

Waël was one of the more active—and talkative—participants in the project in the early days. In addition to serving on the Education workstream, he also participated in Faiths, the weekly all-hands meetings, and some of the special events put on by the Faiths group. He brought the voice of youth and of optimism, and a strong desire to change the way the world works.

He also brought relevant real-world experience because he had lived in a war zone and struggled with Syria's poor public education system. Growing up in Syria, he had an idealistic dream of what it would be like to live in the West. Now he understood that some of the pillars of Western society were crumbling. "I grew up in a dictatorship. Now I live in a democracy, but many people don't have what they need," he said "I'm not being bombed anymore but I see so many challenges, so many negatives in our society. I want to look for solutions that deal with these challenges."

Even though Waël had suffered some bubble-bursting disappointments since arriving in Europe, he remained an idealist. He and his wife, Anais Fourgeaud, had joined a group of like-minded people who planned on someday purchasing a farm near Bordeaux and establishing a communal village where they would grow some of their own food, generate electricity with solar and wind power, and live sustainably. Already, he lived by his values. He mainly bought used clothing and local, organic food.

Waël was in no hurry to pick a professional path. He planned on writing his senior thesis and applying for a master's degree program in some aspect of the humanities. Meanwhile, he was busy learning French to go along with his Arabic, English, German, a bit of Japanese and a bit of Russian. He hoped to travel the world someday.

He thought often of his family back in Aleppo. His parents were still suffering and in danger. He volunteered to help refugees wherever he lived.

Waël had a tattoo on his left forearm of a feather pen with small birds flying above it. He got it soon after he ended his refugee journey in Berlin. The tattoo symbolizes some of his core beliefs: "Instead of using a gun, the pen is the best way to create a revolution," he said. "The birds are about freedom and they also remind me that we are all immigrants in some way."

4

Struggles

When Pivot Projects launched, a key element of the plan was to get funding from foundations to pay for staffing, website expenses, engaging with cities, and technology. Peter Head had set a budget of $400,000 for the first nine months. He was optimistic that the project would be able to get funding—and quickly. Its work was so timely. He knew a lot of people. Funders were signaling that they wanted to pitch in during the crisis.

Each week in the early days, Peter would provide a status report about funding on the all-hands Zoom call. In mid-May, there were no fewer than eight funding prospects totaling $600,000. They included corporate foundations, government-backed funders, and private foundations. He even heard that his own professional association, the Institution of Civil Engineers, was making some COVID-19-related grants. That one seemed like a lock.

Yet with each passing week, one potential funder after another either failed to engage, hadn't decided yet, or turned down Pivot Projects requests for funds. It was frustrating. "We always do a dance on funding. Sometimes we're too ambitions. Other times we get turned down because we're not doing deep research," Peter said. "The problem is there really aren't funding streams for mobilization, collaboration, and systems thinking. The kind of things we're doing, the systems work and connecting things—nobody else is doing it." By midsummer, still no money.

Funding was just one of the challenges that Pivot Projects faced as it ramped up its activities and begin to produce systems maps and reports.

The other high hurdles were running an all-volunteer organization, finding cities that wanted to participate in the project, and getting the attention of policymakers. While Peter and his small team at Ecological Sequestration Trust and Resilience Brokers had experience with these issues—and scars to show for it—their Pivot Projects colleagues with business or academic backgrounds were strangers in a very strange land. The fact that nobody had ever tried to pull off something quite like this before was exciting, but that meant there was no guidebook for doing it, and funders and others in the social establishment apparently didn't know quite what to make of it.

Managing the Unmanageable

The brunt of managing this unwieldy enterprise on a day-to-day basis fell to Colin Harrison. As I noted previously, the leaders decided early to get much of the work done within more than twenty workstreams. They gave the subgroups general instructions: identify the 100 key words, make a Kumu spreadsheet and map, write an interim report, engage with the artificial intelligence (AI) machine, and then engage with cities or regions. But the leaders left it to the groups to figure out what to focus on and how to get their work done. Colin attended many of the workstream meetings and sent countless emails answering questions and clarifying the core management group's intentions.

The biggest issue Colin faced was the fact that the work was voluntary: not only for involvement in Pivot Projects overall but also for the workstreams and leading the workstreams. People chose the workstreams they wanted to be involved with—or, sometimes, none at all. If they didn't like what they saw after a few workstream meetings, they quietly disappeared. Colin and Peter recruited a number of people to be workstream leaders. Some stuck. Others didn't.

Some of the seemingly critical workstreams were slow to get off the ground. These included Climate Change Risks; COVID Impact and Public Health; Housing, Construction, and Land Use; and Jobs, Skills, and Incomes. Housing, Construction, and Land Use, for example, didn't really get going until late July 2020. At that point, Aneta Popiel, a young Polish woman working in London, took the leadership role. She was a manager for a Slovakia-based development company that operates throughout Europe. On a personal level, she was committed to sustainable real estate development and was pressing her company to build greener. She did not procrastinate. Before the workstream's second meeting, she had already written a draft report describing its range of topics and goals. At the second meeting, she assigned each participant to produce a list of key words the group should consider for its system map. This workstream had lingered on the launch pad for weeks; now it was taking off like a rocket.

Another workstream, Industries and Business, had taken a surprising turn right out of the gate. This one was led by Gerard Mooney and Jim Cortada, two former IBMers. The rest of the roster was mostly former or current IBMers as well. During the group's initial meetings, they discussed how important public–private partnerships would be for managing economic stimulus investments in the post-COVID-19 world. They also saw the need for a wide range of stakeholders, from business and government leaders to community members, to designers and architects, working together collaboratively—something they called design charrettes and Resilience Brokers called collaboratories. Instead of creating models of industries and businesses, which had been their assignment, they pursued the development of an engagement model that Pivot Projects could use when it began to work with cities and regions.

Colin sent an email to the other leaders in early summer 2020 listing a handful of matters that he said he was "losing sleep over." One was that some of the humanities-oriented groups were struggling with systems thinking and Kumu mapping. Faiths was a major concern. Here's part of what he wrote:

> The Faiths workstream from the onset has been highly active and has been holding deep and meaningful debates that I have greatly enjoyed. But it shows no signs of converging on a coherent set of thoughts

> relevant to the Pivot Projects study. This is not a criticism of any participants, but rather, I feel, a problem inherent in the intangible nature of faiths and in the way that philosophers and theologians think. A couple of people have suggested asking this workstream to consider the values espoused by the major religions (and Humanitarianism?) and then assessing outcomes from the holistic system against these. This seems plausible, but leads to other problems: How does a computer assess such outcomes?

One Friday afternoon during a one-on-one Zoom call with me, Colin rubbed his eyes and confessed that he was exhausted. It had been a long week—both out in the world and within Pivot Projects. "The world is pretty full of chaos right now," Colin commented. At the time, he was seventy-three years old. Some days, he spent ten hours on Zoom calls. Yet, for months he didn't back off. In many of the workstream meetings, Colin was a calming influence in the background—stepping in from time to time to answer questions or to offer personal observations but never stepping on anybody's toes.

Part of the management challenge within Pivot Projects was the fact that it was so complex and so unlike other projects that the volunteers had worked on previously. Even though detailed information about the group and its activities and a "manual" for getting things done were available in Trello, being involved for many participants was something akin to the cautionary tale of the blind men and the elephant. If they came into the project by attending a workstream meeting first, they might have little understanding of the overall structure and goals. If their first contact was a webinar where indigenous women spoke to the group, who knows what they would think about Pivot Projects. The key thing was for them to attend the Friday all-hands meetings, where they could see the bigger picture.

After many of the workstreams completed their initial reports, systems models, and Kumu diagrams in July 2020, there was an initial sense of elation. They had splashed some paint on the canvas. But almost immediately, there was a bit of a letdown for some people. What was next? Why weren't the leaders more directive? During one of the workstream Zoom calls, two people were on by themselves at first, and they began venting their frustrations.

> SPEAKER ONE: Nobody's being really clear about what's happening next. There's a lot of talk about trying to figure that out.
> SPEAKER TWO: The project leaders need to understand what the focus is and tell us what the focus is here. We've been thrashing around, treading water. I'm afraid that people may start to lose interest and drift off. It would be a sad waste of an opportunity.

When Colin got wind of the air of uncertainty, he was a bit irritated. He had spelled out next steps in a recent newsletter, but it seemed like even some of the most committed volunteers hadn't read it. At the all-hands meeting on August 1, he was uncharacteristically blunt. "This is not even the bloody beginning," he said. "We didn't come here to make Kumu graphs and write preliminary findings. We came to change the bloody world. *That's* what comes next. There's a lot to be done." He put up a slide listing the next set of tasks covering the technical aspects of the project. Here's a streamlined version:

- Complete the Kumu knowledge graphs, reports, datasets, and references.
- Test and then integrate the Kumu knowledge graphs with one another.
- Load Pivot Projects systems models into SparkBeyond's AI machine.
- Check for any human bias in the AI platform so that we can build trust all over the world.
- Set up filters to sift through the vast numbers of hypotheses generated and to find the "best" ideas based on our principles and priorities.
- Configure the AI platform for specific cities and regions and for the issues that concern them.
- Train people internally and externally to use the tools. Volunteers then move out of workstreams to help cities and regions to make change happen.

Wooing Cities

A few days later, Peter Head followed up with a memo of his own describing other tasks that lay ahead. Initially, the group had posted some information about Pivot Projects on the Resilience Brokers website, but now they planned on launching their own website and trying to drive a high volume of traffic to it via social media. The web-hosting company, Wix.com, had agreed to set up the site and host it on a pro bono basis. (Thanks, Wix.com!) There would be a crowd-funding feature on the website, so they hoped that they would be able to raise some money. Another key initiative would be to systematically reach out to cities around the world to try to set up pilot projects where the modeling and AI tools could be tried.

Difficulty engaging with cities was another source of frustration for Peter. Originally, he had set a target of engaging with about fifteen cities and regions. The most likely prospects were members of some of the global sustainable cities networks, including the C40, a global community of more than 200 city, state, and regional governments committed to doing their part to reduce carbon emissions and keep global warming under 2°C (3.6°F). In the early days of Pivot Projects, just like with the potential funding sources, Peter reported weekly on the status of outreach to cities. A number of places emerged as potential targets, including Exeter in the United Kingdom; Cape Town in South Africa; Oaxaca in Mexico; Singapore; and Portland, Oregon, in the United States. But Pivot Projects didn't yet have the staffing or money it needed to make contact and engage with cities.

Resilience Brokers had extensive experience in working with cities and regions—helping them use data and systems models to understand their problems better so they could create effective solutions. The engagement in the metropolitan area of Accra, Ghana, with 4 million people, was the most extensive. Funded by an agency of the UK government, the Department for International Development, Resilience Brokers convened a meeting of a group of government and business leaders, sustainability experts, and community leaders to discuss their challenges and choose the most urgent issue to be targeted, which turned out to be water and sanitation. The collaboratory brought together many different kinds of people and many disciplines, and provided a means for them to solve problems together.

The second part of the project was creating the technology platform, which they called Resilience.io. Think of the computer game SimCity, where a player develops a city from undeveloped land and adds buildings, roads, schools, and hospitals. The player levies taxes, creates a budget, and sets policies and regulations. All the while, as a tract of virtual land grows into a megalopolis, human avatars called "Sims" live, work, and play. The player wins if she or he is able to build and operate a city while keeping the Sims happy and productive. Resilience.io was similar to SimCity, except that it was a model of a real city or region, and, the people using it were looking at real challenges like unemployment, water pollution, and flooding—and the real Sims were frequently angry.

The idea was that participants in the Accra collaboratory would pull together all the available studies and data addressing water and sanitation in Accra and in general, they would build models of the relevant systems, and then they would use software to spot interdependencies and work up promising investment strategies. The technology platform was built with the Java software programming language and open-source software components to minimize the costs to users. They used two kinds of modeling. The first was dynamic, agent-based modeling. The population was represented by agents, embodied in the model, so that the demand for water and sanitation across the communities could be modeled across each day, week, and month.

They also used a then-brand-new modeling approach called resource technology network modeling. This approach measured the inputs and outputs of water supply, water treatment, and sewage collection systems as well as the related energy, materials, and economic aspects of those processes. The existing inadequate supply systems were modeled first. The team then used the model to simulate the direct and indirect effects of different investment strategies.

The Resilience.io platform worked for Accra. The collaboratory used it to develop improvements to the water and waste systems, including treatment plants, pipelines, and sanitation facilities. The work then led to a master plan being created for expanding on the initial improvements and demonstrating costs and benefits. It showed that systems modeling and multistakeholder collaboration worked. But it included a relatively narrow spectrum of systems—those directly related to water and wastewater. To Peter, Pivot Projects represented an opportunity to take the early work done in Ghana and go global.

By midsummer 2020, however, none of the cities Pivot Projects considered prospects for collaboration had advanced past initial discussions, partly because they were so busy dealing with the spread of COVID-19 and the collapse, to one extent or another, of their economies. In Cape Town, for instance, there had been alarming increases in crime, poverty, unemployment, food insecurity, and domestic violence.[1] But Peter said that the biggest hurdle was always having money to pay for engagement work. He said that most cities are short of cash and cannot free up people's time, even for something worthwhile. "Everybody in the world is really stressed. The localities don't have money. We're all crumbling a bit, I reckon," he said.

Influencing Policymakers

Another challenge was breaking through and influencing policymakers—at the UN, at the European Union, and in national governments. These folks had years of thick reports about the dangers of climate change piling up on their shelves. Now, they were making decisions about how to stimulate economies and respond to humanitarian crises around the world. They were deluged with proposals, criticism, and reports.

In his career, Peter had enjoyed plenty of success in the report-writing category. The highlight was his work on the UN's Sustainable Development Goals (SDGs) and in particular SDG 11 for cities. The goals had set a new agenda for governments and nongovernmental organizations (NGOs) around the world. The report he had created for the Institute of Civil Engineers, "Entering the Ecological Age: The Engineer's

Role," had landed him and his ideas in cover stories published by major news organizations. Another report he had coauthored, "Safeguarding Human Health in the Anthropocene Epoch," which extoled the value of systems thinking, had been cited in other scholarly articles hundreds of times. He felt that his published work had deepened understanding of the importance of the interactions of human and natural systems, and had helped make systems modeling a mainstream approach for social and environmental problem solving. "A lot of people are cynical about reports—questioning whether they're worthwhile. But I believe they're useful, particularly when you engage with people on what they mean to them locally," he said. "Plus, it's better to do something than to sit at home and complain about things."

During the summer, Peter submitted two significant reports to policymakers on behalf of Pivot Projects. One was a paper about drought, land regeneration, and reforestation, which was published as part of a special report on drought for UN meetings in 2021. In it, Peter and the team made the case for policies to reverse the loss of forests by employing regenerative agriculture and forestry techniques. This would also reduce the risks of future pandemics. In the second report, prepared for the Group of 20 (G20) technical conference in late 2020, Peter and his coauthors proposed that the G20 countries create a set of policies and actions to facilitate the use of a data-driven, integrated planning and procurement approach to sustainable infrastructure development.

Peter also contacted the COP26 organizers and offered to support their preparations for the conference, which had been rescheduled from late 2020 to late 2021 because of the COVID-19 pandemic. He was hopeful that the work of Pivot Projects teams would influence policymakers both immediately, as they considered actions in the wake of COVID-19, and for the longer term. He also hoped they would raise the profile of Pivot Projects—and lead to funding.

Peter had stepped back from the day-to-day project work at Resilience Brokers in 2019, and this gave him time to be immersed neck-deep in Pivot Projects. It was taxing work. Like Colin, he was seventy-three years old. He spent more than one week writing the first draft of the UN report connecting climate change and drought and, after he got feedback from others involved, he rewrote the submission. That kept him up working in his study until 3 a.m. the day the report was due.

Not every attempt to reach policymakers was a success. Peter had submitted a bid to the Paris Peace Forum 2020 (organized by the French government), which planned on spotlighting new ideas for dealing with the COVID-19 pandemic. It seemed like a perfect target for Pivot Projects, but he got turned down. Peter's attitude: Too bad; on to the next thing.

Peter, Colin, and other leaders of Pivot Projects were managing a sprawling enterprise. The organization had to be multidexterous: searching for money and for additional volunteers, producing articles and presentations, shepherding the workstreams through their tasks, collaborating with the folks at SparkBeyond to build the ideation platform, and reaching out to find cities that might be willing to try this new approach to decision making. It was like playing three-dimensional chess and spinning plates on sticks at the same time.

Back to Money

Money had always been a sore point for Ecological Sequestration Trust and Resilience Brokers. Peter had launched the trust with money from a cashed-in pension. He hoped to be able to get sustaining funds from foundations, but he had had no luck. Later, one of the reasons for creating Resilience Brokers was the idea that they could sell shares in the company to financial firms or corporations who had pledged to put a portion of their investment funds in enterprises focused on sustainability. They hoped to raise £100 million in equity to fund Resilience Brokers' rollout of support to city regions, paying in part to fund the development of Resilience.io. Once the platform was built, it could have been used in any city or region that needed help simply by adding local data and models to it. Again, there was no luck with funding. "People don't think strategically, and they don't think long term. They just think about one-off projects," said Peter. "It has been a very hard road."

Peter and the organization instead sought individual projects for which funding was available. That approach landed them a waste-to-energy investigation in the Hunter region in Australia, where there was a severe drought and water shortage; in Mongolia, where there was an air quality problem; and in Norway to help cities attract investments for climate compatible infrastructure. "We have gone from hand to mouth, project to project," Peter said. That made it difficult to fund the staff and activities of Resilience Brokers. But it had another deleterious effect: The organization had not been able to build the massive platform for systems modeling it had begun working on in Ghana. Hopefully, Pivot Projects would do that—although the modeling approach and the analytics technologies were different.

Meanwhile, Resilience Brokers was suffering. It was a small organization that operated virtually, with just one small office. It hired experts when it needed them for particular projects. It ran lean, but the same stresses that were crushing businesses and individuals all over the world were roughing up Resilience Brokers as well. Existing projects slowed to a crawl. New projects were hard to come by. Peter was putting almost all of his time and effort into Pivot Projects, leaving Stephen Passmore, Resilience Brokers' CEO, to keep the little craft afloat with day-to-day project work. "We're bailing frantically. We're in a precarious situation," said Stephen. "We have a couple of small contracts for now but we don't know if we have commercial viability into the medium term." He had to put employees on furlough and close the office to save cash.

Stephen fought discouragement daily. He was an optimist by temperament, but the pressures of COVID-19 on the organization were sorely testing him. He had valued Peter's leadership and noted that Peter never seemed to get down—at least publicly. "He once told me that the biggest job of the leader is maintaining hope, but my trust in his ability to do so is waning," Stephen said.

Peter confirmed that he never gets deeply discouraged—even when he's by himself in the wee hours of the morning in his studio writing reports for the UN. "I have absolute belief that what we're doing is right and important," he said. "I might get disappointed by something, but I bounce back. I decide, 'Let's try another way.' "

The Pivot Projects leaders felt elated after Peter's presentation via Zoom to a British government group on August 7, 2020. They had opened a direct channel to the leaders of one of the most important countries in the world. They looked forward to influencing other policymakers in similar ways, including the G20 and ultimately COP26. This saving-the-world stuff was by turns exhilarating and exhausting. But maybe, just maybe, they would actually be able to pull this thing off.

One major concern remained, and it was big: After nearly six months as a group, they hadn't been able to raise money to expand the project, which meant that they couldn't assure SparkBeyond that there would be money to help pay for computing resources, and they weren't sure they would be able to help keep Resilience Brokers afloat. Pivot Projects had benefited from the generosity of a couple of tech companies. I mentioned Wix.com earlier. Slack, the maker of the collaboration platform, allowed Pivot Projects to use it for free. Also, Pivot Projects was using a free version of Kumu.

Peter was frustrated but determined to muscle through. He told the story about advice a friend had once given him. The friend, Tim Smit, was the founder of the Eden Project, an effort to build a complex of geodesic domes in Cornwall, in the southwest corner of Great Britain. Housed in the domes were to be living replicas of the Brazilian rainforest and other biomes. It was to be a place where schoolchildren and people of all ages could visit and learn and hopefully leave inspired and committed to saving nature. Smit had worked for many years with very little money, but in the end he succeeded at turning his vision into reality. He had once told Peter that with little money you just have to move ahead at full speed with the confidence that at some point the money would appear. It's like running full speed at a wall and hoping that somebody opens a door before you crash into it. "That's what we're doing," Peter said. "We're running harder and harder at the wall and saying, 'Please open the door for us.' "

PROFILE

Anh Nguyen

Vietnamese fish exporter studying in Sweden

When Anh Nguyen was a growing up in Vietnam, her father invented a semiautomatic egg incubator that made a huge difference not just for her family but for the entire village, Thượng Yến. The family had a small farm where they raised chickens and ducks, and her father had ideas about how to incubate the eggs on a large scale. The family had no paper or computer, so her father drew designs with chalk on the brick floor of their home. Ultimately, his work bore fruit: he designed an incubation system that was capable of hatching 15,000 eggs at a time and at a cost that was just 10 percent of imported systems. He allowed anyone in the village to copy the design. "The machine became famous. He made the village rich," said Anh.

Watching her father design something that could transform the economic fortunes of a whole village inspired Anh to become an entrepreneur and to help others after she became an adult. Today, she runs a salmon exporting business in Norway. She's also studying for a master's degree in multiculturalism.

Anh had another formative experience in her youth that shaped her personality. When she was very young, before the incubator came along, she lived for a few years with her grandparents in another village, an hour's drive from her parents. During her first years in elementary school, she would get up before dawn and work in the fields before going to classes. She carried water from a pond to the garden, and she sold vegetables in the village market. The family didn't have the money to use pesticides to ward off insects, so Anh would pick them off the vegetables by hand. She captured grasshoppers in a jar and her grandmother would cook them for dinner.

These experiences from her youth are what attracted her to Pivot Projects. She learned about the initiative from a friend in Norway who was a friend of Peter Head. The idea of working with a group of people to develop ways of living that are more natural and sustainable appealed to her. She joined and became a member of the Food and Agriculture workstream. Also, after she learned that the group needed more young people from around the world, she put out the word on Facebook, and because she had a lot of friends in a lot of countries, her appeal attracted more than 100 young people. When she learned that one of the young volunteers who lived in Nepal didn't have a laptop computer, she shipped him one that she didn't use anymore.

Anh initially struggled with the research. The project was so complex that she could not figure out how to contribute as much as she wanted to. She almost quit in frustration very early but was persuaded to stay on by Colin Harrison. Gradually, she gained confidence.

She was helped by the writings of a famous designer, Damien Newman, an American who had developed his craft while working for the Silicon Valley industrial design firm IDEO. During his years of working with clients, Damien had observed that many expected the design process to be linear and simple. So, early in engagements, he had to teach them otherwise so they could understand what he and his team were doing and thus the clients could participate constructively. He invented a diagram called The Design Squiggle, which he used to illustrate the design process. It conveys the feeling of the journey from researching,

Noise / Uncertainty / Pattern / Insights

Clarity / Focus

Research and Synthesis

Concept / Prototype

Design

The Design Squiggle process

"The Process of The Design Squiggle" by Damien Newman, thedesignsquiggle.com

uncovering insights, generating concepts and prototypes, and eventually landing on a final design.

When the Food and Agriculture workstream started to meet, it was not clear to Anh what it would accomplish and how it would do so. The world was in chaos, and, in her view, so was the workstream to some extent. She understood that designing solutions to the world's food problems was in some says like designing a new vacuum cleaner or medical imaging machine—only more complex. So she let Damien Newman's Design Squiggle be her guide. "We started with a mess," Anh said. "It was like walking in the dark in a tunnel believing that there will be a door at the end. We are doing this with belief and hope. I don't feel like I'm lost anymore. I believe we are progressing."

The Food and Agriculture group eventually produced a report, a Kumu spreadsheet, and a list of questions for the artificial intelligence (AI) machine. Along the way, Anh got some ideas that might be useful in her salmon exporting business, which had been temporarily shut down in the early days of the COVID-19 pandemic. She investigated the idea of raising salmon in containers built into a cargo ship. The ship would move from port to port, offloading salmon when they were mature and ready for sale. In this way, production and distribution would be combined—saving money and energy.

By midsummer, Anh felt comfortable with the Pivot Projects process and her role in it. "I'm not scared of uncertainty anymore," she said.

5

Remapping the World

During one of the earliest all-hands Zoom meetings for Pivot Projects, Andre Head, son of Peter Head and the group's collaboration technology maven, shared his computer screen with the group. Everyone saw an array of colored dots—pink, orange, and burgundy—with swoopy black lines connecting them to one another. Andre informed his colleagues that they were looking at a Kumu diagram of Pivot Projects called the Member Network Map. He explained that the orange dots in the middle of the map were the twenty-plus workstreams. The pink dots, also in the middle, represented core team members. Burgundy dots, mostly small dots around the edges of the map, represented the regular members. The swoopy lines connected people with workstreams, so you could see clusters of people dots connected to workstream dots. The dots were larger for people who were connected to a number of workstreams, and larger for workstreams that had more members. Most people on the Zoom call were muted, but several people oohed and aahed audibly.

Then Andre placed his thumb and index finger on the track pad on his computer and spread them apart, instantly expanding the size of the map and showing close up the pink dot representing Colin Harrison, who ran day-to-day operations. You could see that this was the largest of the pink dots, and you understood that this was because Colin was participating in all of the workstreams that were running at that point. Then Andre clicked on one of the orange dots, which caused all of the

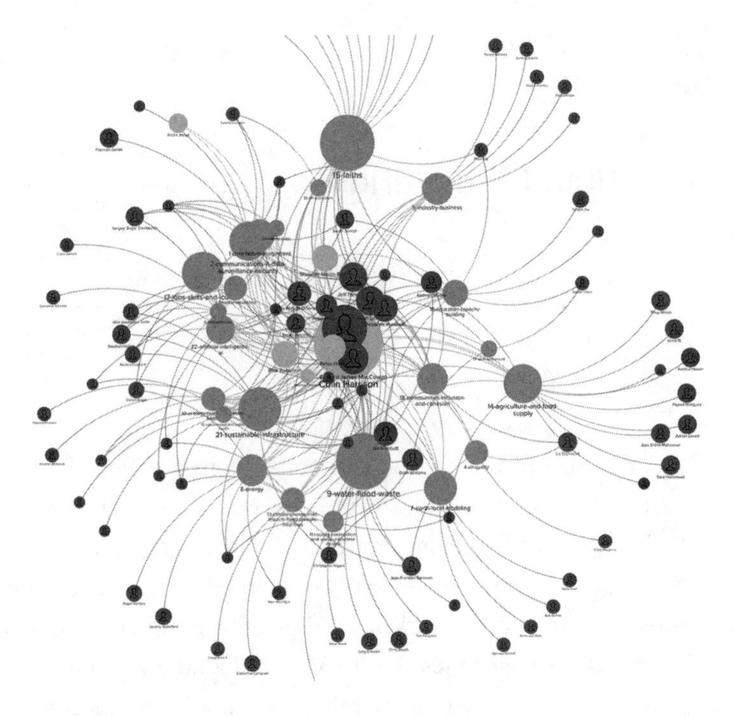

The Member Network Map, a Kumu diagram of the Pivot Projects

other workstream dots and the dots for people who were not connected to this workstream to fade so that they were just barely visible. You could see clearly that single workstream and the people who participated in it. Swoopy lines coming from all over the map connected people dots with the highlighted workstream dot. It seems likely that several people on the call, and perhaps the majority, now fully understood the structure of Pivot Projects for the first time.

This is the magic of Kumu, a computer visualization tool that enables people to better express or understand complex relationships between things. There are many computer visualization tools in the world, but this one is immensely popular for a number of reasons. It's relatively easy to learn and use—although it is not intuitive. It's extremely flexible: The dots can be used to represent things, words, and concepts. And it scales up and down: you can pull back and get a comprehensive view of a large constellation of dots and swoopy lines, or you can pull in close and see a small piece of the picture.

Andre had created the Member Network Map by linking it to the group's Slack communications application. When participants joined the group Slack account, they became dots on the Member Network Map. In Slack, members could join workstreams by clicking on digital buttons. That created the swoopy lines between individual members and the workstream dots in Kumu.

Building Ontological Models

While the Member Network Map was useful for providing real-time snapshots of how Pivot Projects was organized and operated, the main job for Kumu within the organization was something quite different. The tool was being used to help the workstreams build ontological maps or models of their topics of interest—be they public health, climate change, economics, construction, education, water, or faiths. (In the lingo of information science, an ontology is a representation and naming of the concepts, data, and entities within a system or domain.)

As the basis for its ontologies, Pivot Projects used the definitions that had been collected by Wikipedia, the free, online encyclopedia that anyone can use and edit. They chose Wikipedia as their source of definitions for two reasons: it is a nearly universally accepted source of definitions and information, and SparkBeyond uses Wikipedia as the basis for the knowledge and the reasoning in its artificial intelligence (AI) machine. An ontology built in Kumu based on Wikipedia definitions and content could communicate directly with the AI.

The Pivot Projects teams were using Kumu to build models of the systems they were studying, humanmade and natural, and how those entities interacted with one another. By building these models, and later by importing the models into the computer, they would be able to get the computer's help in reasoning about their models—understanding how they worked and predicting how they might behave differently if something new was introduced into the model or the model was changed.

Members of the Pivot Projects workstreams were to begin by choosing about 100 key words in each domain, which were represented by the dots and were called nodes—in Kumu speak. The swoopy lines connecting the dots were called edges. They showed relationships between nodes. The teams created the Kumu maps by first creating spreadsheets that contained and organized all of the entities in their maps and the links between them. All of the nodes linked to pages in Wikipedia.

Kumu was created in 2011 by two brothers living in Hawaii, Jeff and Ryan Mohr. They were working with the Omidyar Network, a group of nongovernmental organizations (NGOs) and social businesses backed by eBay cofounder and philanthropist Pierre Omidyar. The brothers designed Kumu to help the organizations assess the impact of their work and develop more powerful ways of working together. Kumu made it easier to see potential connections. But the Mohr brothers saw that the visualization tool could be applied to a wide variety of thinking and collaboration tasks—so they created the company Kumu and improved the technology. Today, the Kumu tool is used for several different thinking tasks, including systems mapping, which helps people understand complex systems; stakeholder mapping, which explores the web of interests and relationships among a group of people; social network mapping, which reveals the connections and the key players in informal networks; and concept mapping, which helps groups brainstorm ideas.

The Kumu Guru

Andre Head introduced the group to Kumu, but it was Ian Abbott-Donnelly who became the group's Kumu guru. Ian was a member of the so-called IBM mafia. He had worked on the Smarter Planet team, but he retired early so he could have the freedom to explore a wide variety of pursuits, ranging from road bicycling to painting on canvas, to collaborative research. He lived in Stamford, UK, a village about thirty miles north of Cambridge University, where he was an associate of the University's Centre for Science and Policy.

Ian Abbott-Donnelly

Ian had been trained as a seismologist, but he was always intrigued by computers. At each stage of his career, he was typically one of the first people within his group to try a new computing technology, and so it was with Kumu. Nobody assigned the job of mastering Kumu to Ian. He just saw that somebody should do it and he took it on. That's the way most things got done. While he insisted that he was only one step ahead of others in Pivot Projects, in fact he was the trailblazer who quickly understood the value of the technology and learned how to use it. "It was a conscious process of learning by doing," Ian explained. "You click on everything and see what it does. You read a bit of the manual. Then you start building a map."

You might say Ian was predisposed to understand Kumu. In the 1990s, he had worked as a technologist for Thomas Cook, which was then the largest travel agency in the world. At the time, the travel agency industry was beginning the process of migrating online, but there was still plenty of demand for physical offices and agents; they just had to become much more efficient. To learn how things worked on the retail level, Ian would visit Thomas Cook offices and pretend to read the magazines in the waiting areas. In fact, he was observing to see where bottlenecks developed and queues formed, which frustrated customers. He drew illustrations on paper of the processes the offices used to get work done.

His mapped-out observations were simple models of systems. They helped him and others in the organization make the argument that telephone inquiries should be shifted from the retail offices to call centers so agents could focus on the walk-ins and keep them happy.

That was a very simple example of using visualization to understand a system and diagnose its faults. At IBM, Ian learned about complex adaptive systems and systems thinking. These were essential elements of IBM's Smarter Cities strategy—making cities, transportation systems, water systems, and the like, work better. I'll explain systems thinking in detail in chapter 6. For now, suffice it to say that Ian's understanding of how systems act and interact helped him lead Pivot Projects' foray into Kumu mapping.

One of Ian's first little Kumu experiments came when his mother broke the spine on her address book. He offered to transfer the names, addresses, and phone numbers into a spreadsheet, which he would print out for her. But he also saw the opportunity to transform the information into a Kumu map. By color-coding people based on their surnames and on their locations, he was able to see who was related to whom and where there were clusters of people living close together.

To learn more, Ian joined the Water and Information Technology workstreams. These were led by former IBMers—Peter Williams for Water and Rick Robinson for Information Technology. Both groups got started rapidly and were efficient. They began amassing their key words, and, with Ian's help, they began building Kumu maps. "I realized we were making system influence diagrams," Ian said. "You'd see that this node influences that one, and that one influences this other one. Click. Drag. Click. Drag. Click. Drag. And you've got it done."

Anticipating that it would be useful to track the status of each workstream's Kumu work, Ian set up a spreadsheet with all twenty workstreams on it, and he touched base with the leaders from time to time to update their status. He enlisted one of the younger Pivot Projects volunteers, Andre Hamm (one of my nephews), a software programmer living in Salt Lake City, Utah, to prepare a Kumu crib sheet for the workstream leaders. Andre also wrote a little software program to automate loading Wikipedia URLs into Kumu.

Putting Kumu to Work

With Ian's guidance, the workstreams began choosing key words, creating spreadsheets, connecting nodes to one another with edges, and visualizing their systems on Kumu. Because the models would eventually be loaded into SparkBeyond's AI machine, they had to be organized in a certain way. Each node would be categorized within the spreadsheet as a resource, a process, or an attribute.

At some point, all of the systems created by the workstreams would be connected and rolled up into one big spreadsheet and one big Kumu map. Most of the linkages, or edges, would be lateral, but Ian also realized that there would be some higher-level nodes that many of the other nodes linked to. Ultimately, Kumu would function essentially like a three-dimensional model. Some of these higher-level nodes would be resources, such as the natural environment, society, and the human-built environment. Others would be processes, such as climate change and urban planning. Because Pivot Projects was advocating significant changes in the way society operates, some very important high-level processes would come into play, among them transformation and regeneration.

As time went on, the teams would add more resources, more processes, and more attributes to their spreadsheets and Kumu maps. They typically started with big, broad concepts, but gradually they zeroed in on the issues they wanted to explore with the SparkBeyond machine. The Information Technology workstream, for instance, focused primarily on the digital economy because its members knew that online communications, commerce, and collaboration were going to be key transformational elements in the post-COVID-19 world.

The workstreams didn't have to do all of this mapping work themselves because one of the essential principles of Kumu, the company, is sharing. It is in some ways like open-source software. In open-source software, development projects are typically done collaboratively—often with contributions from a wide range of organizations and individuals. Anyone who wants to can use, change, and distribute the software as long as they follow the terms of a license. In the Kumu world, individuals and organizations can choose to make diagrams for free and those diagrams can be viewed by anyone—and can be incorporated into other Kumu diagrams. For instance, a number of Kumu maps relating to the UN's Sustainable Development Goals (SDGs) had already been created. Pivot Projects could pop elements of those diagrams into its whole-Earth model.

You can begin to see the potential power of Kumu in the hands of the Pivot Projects workstreams. Just as cartographers had developed more detailed and accurate maps of the world, Ian and his colleagues were using Kumu to map the world of things, ideas, processes, and relationships in ways that, as far as they knew, had never been attempted before.

Wrestling with Complexity

The workstreams that focused primarily on science and engineering had an easy time of it. For example, it was clear that water was a resource, water treatment was a process, and the type of treatment involved was an attribute. But Kumu was a challenge for some of the groups that were focused on the social sciences and the humanities.

The Faiths workstream faced some of the most daunting issues. As I mentioned in chapter 3, the list of key words the group generated changed wildly up and down in the early weeks of the project. Initially, each member came up with her or his own list of suggested key words, which resulted in a lot of free association, recalled Sarah Joseph, a member of the group who had an affinity for technology. The suggestions, as she recalled them, included air, water, sun, moon, dancing, and singing. "We had all these words and couldn't make heads or tails of them," Sarah said. The group made a mistake that's probably not unusual for the kind of people they are—they overthought their task. At one point, in addition to categorizing each key word as a resource, process, or attribute, they organized them within four other categories: self, other, universe, and ineffability.

The task of making a truly effective Kumu map fell mainly to Sarah; Waël Alafandi, a Syrian refugee; and Shulamit Morris-Evans, the Extinction Rebellion activist. Sarah had been the editor of a magazine for a number of years, so she knew how to get stuff done on a deadline. The group ultimately decided to make two Kumu maps: one profiling organized religions and the other looking more broadly at spirituality, beliefs, and ethics. By late August 2020, they had 200 nodes and 900 edges—and the numbers were still climbing.

The Education workstream also had its share of Kumu challenges. The group decided early on that they would focus primarily on three issues in education: teaching children about climate change, preparing them for the occupations that will emerge in a more sustainable society, and addressing well-being and mental health issues among students and teachers alike. But they agreed more broadly that one of the biggest problems in education was that boards of education, administrators, and teachers decide what young people needed to learn based on the world as it is today rather than reimagining the world of the future and helping them prepare to live in it. Another critical factor was granting young people agency. "How do we find ways to listen and discover what *they* want and what *they* need?" said the workstream leader, Alan Dean. He ran an organization in the United Kingdom called Burning-2Learn, which develops programs aimed at raising self-esteem and confidence for students. "We want to educate them for tomorrow, but, since COVID-19, we don't know what tomorrow will be," Alan said. "So, for me, it's all about listening."

As with the Faiths group, Education took a while to figure out Kumu. Unlike the Faiths group, their problem was with the internal dynamics of the Kumu tool rather than with organizing their thoughts and priorities. Most of them didn't have much experience with spreadsheets. They started with a group brainstorming session to identify their initial batch of key words, then refined them. Bill McKenna, who has a doctorate in science, technology, engineering, and math (STEM) education from the University of Texas, Austin, became the group's Kumu maven. On review of the first selections, he noticed that most of the words they had chosen were positive and connected to well-being: including *belonging*, *engagement*, *friendship*, and *creativity*. That didn't seem appropriate in a world beset with woes, so they added some negatives: *anxiety*, *stress*, *self-harm*, *suicide*, and *depression*.

When the Education workstream started making their Kumu map, they had selected their topics of interest and wanted to focus on the concepts that would help them explore and develop them. Only weeks later did they notice that they had left out many of the traditional key words normally associated with education—such as *math*, *science*, and *history*. Bill figured that was okay. Their job was to rethink education and to set off in a new direction. "It's in our genes that school is about reading, writing, and arithmetic. Well, kinda," McKenna said. "We didn't leave the words out on purpose, but I'm proud of us."

One of the main challenges that all of the groups faced was that nothing like this had been tried before, so it wasn't clear how best to build the Kumu models. They knew they couldn't—and didn't need to—build extensive conventional models of each of the systems of interest. Well-understood systems were already encoded in Wikipedia, which was essentially an encyclopedia of human knowledge. What more should they bring to SparkBeyond's AI machine? What did it need from them if humans and machine were to have a fruitful discussion? The groups were like the crew on a ship at sea trying to find a safe harbor on a foggy day.

Learning by Doing

It took longer than the group had hoped to be able to feed the Kumu spreadsheets into SparkBeyond's AI platform. The Israeli company was still in the process of building one of its key applications, called Research Studio, which Shay Hershkovitz, the SparkBeyond vice president, saw as the primary tool that Pivot Projects would use. The core technology was in place, but software engineers were still working on the user interface and navigability. Fortunately, that didn't stop Ian and a handful of others from experimenting with the technology. They needed a system to focus on, and after casting around for a bit, they settled on cities and their infrastructure.

Peter Head had posted an article on Slack a few weeks earlier comparing the job creation potential for various kinds of infrastructure investments. It showed that big infrastructure projects sometimes don't produce a wealth of long-term jobs. For instance, the High Speed 2 rail line from London to Manchester and Leeds in the United Kingdom was expected to produce only two long-term jobs per £1 million spent. Meanwhile, the article reported, renewable energy investments, such as installing solar energy panel arrays on rooftops and in fields, created about fifteen long-term jobs per £1 million spent. Government officials and engineers typically focus on the construction jobs created by a large infrastructure project rather than on the longer-term impacts. Peter believed that was shortsighted, especially given the importance of providing people with occupations that will be relevant and useful in the looming battle against climate change.

The prospect of stimulating economies, building sustainable infrastructure, and producing future-oriented jobs was a win-win-win for everybody. Ian gave the project a name: the Cities-Sustainable Infrastructure-Jobs model. These were critically important issues. It was time to get going on the next phase of Kumu development. To get the job done quickly, Ian recruited a couple of young volunteers to help.

The first was James Green, a recent graduate of the University of Brighton in the United Kingdom. He had a master's degree in environmental policy. He was living at home with his parents and working in a record store, Vinyl Hunter, in his hometown of Bury St. Edmunds. He came to Pivot Projects via a roundabout route. He had posted on Facebook that he wanted to buy tickets for a sold-out show by the psychedelic rock band Toy in nearby Suffolk. He struck up a conversation with the guy who sold him the tickets—"moaning about my depressing, jobless, penniless, living-at-home situation," James recalled later. The next thing he knew, the guy introduced him to Stephen Passmore, CEO of Resilience Brokers and a central character in Pivot Projects. They spoke and soon James was in. (He later joined the core leadership group.)

The second recruit was Andre Hamm, the young software programmer who had written the Kumu crib sheet. Andre worked for a software company in Salt Lake City that uses natural language processing to provide real-time captioning during phone calls for people with hearing impairments. He had recently returned from a whitewater rafting trip on the San Juan River in southern Utah. Andre was an outdoorsman and a lover of nature. He was ready to plunge back into the project.

Ian's plan was to add information about cities and job creation to the Kumu map that the Sustainable Infrastructure group had built. It would be a mash-up of three substantial socioeconomic systems. For whatever reason, Pivot Projects' leaders had not seen the need for creating a workgroup focusing specifically on the urban environment, although other groups covered issues that are important to cities. The workstream for Jobs, Skills, and Incomes had never gotten off the ground. This was an opportunity to fill those gaps in the overall model and create a useful tool for government leaders at the same time.

The team went at it. They were unencumbered by the unruly discussions and searches for direction that had slowed progress for some of the other workgroups. On the other hand, they missed some of the benefits that come from a diverse group of people wrestling with complex ideas. It would be an experiment in how best to get this sort of thing done. They built the model in a couple of weeks. Working fast was important, but it was more important that they were learning by doing.

The Big Kumu

I recall years ago reading a quote from an American author who said writing a novel was like walking across Asia on your knees. It seems to me that the Pivot Projects journey to creating its first big Kumu map was something similar—only for a group rather than one brave individual. This months-long process was truly taxing: all the brainstorming, the thrashing about, the debating over what should be included and what should be left out. I can tell you as one who has helped make a Kumu map, in my case triangulating the topics "rivers," "arts," and "education," that the effort to identify and connect the key concepts in an area of interest *does* help focus the mind.

When they were finished with the initial Kumu, it consisted of 1,642 nodes and 5,020 edges—the lines connecting the dots and concepts to one another. Viewed on a computer screen, it looked like a giant blue tumbleweed. The map gave shape to Pivot Projects' areas of interest, the twenty systems that the workstreams had focused on. But heir selective worldview was also reinforced and enriched by the entirety of the English version of Wikipedia, 3.9 billion words wrestled over for years by millions of volunteer contributors and editors.

One day at an all-hands meeting, Ian shared the blue tumbleweed with the assembled group. He turned the diagram this way and that, and zoomed in and out. The Kumu was awe inspiring. "This map represents the collective intelligence that's inside your brains. It's a manageable, human-readable view of the world," Ian said. "With this map, you can explore and make sense of something that's quite complex. You can think about it in a systemic way."

All I could think was, "Wow."

PROFILE

Tom Rossiter

American architect and photographer

In November 2019, a couple of months before COVID-19 spread globally, Tom Rossiter built a steel platform on the roof of his home in Chicago for shooting photos of the city's skyline. A professional photographer, he set out to make a year-long photographic meditation. He set up a fifty-megapixel camera so he could capture high-resolution images. One night after the COVID-19 pandemic started, he was sitting on the roof platform and he had a revelation: What would happen if we turned all the lights out? He realized that we would likely see the Milky Way, our window on the universe.

He decided to make composite photographs that would combine images of the skyline of Chicago with images of the Milky Way. He wanted to capture the idea of reflection, the way sky and mountains reflect on the surface of a mountain lake, but instead of using a reflection on water, he would use reverse images of the skyline and the night sky. This concept became one of the foundations of his project, "Anthropocene Mountain," which evolved over time into a 120-minute video that he called a meditation on humanity's relationship with planet Earth—and the need to change the way we live. It was a slowly-evolving series of video images capturing everything from forest fires and waterfalls to people in motion and the spread of the coronavirus.

Tom saw city skylines as something akin to mountain ranges—only built by humans rather than natural forces. With the photos, "I wanted to look at the city abstractly, as it sits in the universe, and as a system of human occupancy," Tom said.

It was Peter Head who invited Tom to join Pivot Projects and lead the Arts and Culture workstream. They knew each other from working together at AECOM, a global engineering and architecture firm. They each ran divisions of the company. They would see each other at board meetings in London and had become friends. "The rest of the people at the table were really interested in EBITDA [an accounting acronym for earnings before interest, taxes, depreciation, and amortization]; we liked to talk about art, sustainability, and native cultures," recalled Tom.

Tom had come to his spot in the universe via a curvy route. He grew up a surfer kid in southern California, worked as a potter, studied architecture at the Rhode Island School of Design, interned with the influential polymath designers Ray and Charles Eames, worked as an architect, cofounded a successful architecture firm what was bought by AECOM, and later became a photographer specializing in architecture—creating glossy spreads that help architects win awards for their designs.

In Pivot Projects, he led a lively group of people through the process of picking key words, building a spreadsheet, and producing a Kumu diagram. But Kumu wasn't his thing. His real passion was combining his "Anthropocene Mountain" project with Pivot Projects. He wanted to bring storytelling to the enterprise via still photography, videos, and music. He believed that convincing a large number of people to put society on a sustainable path could not be accomplished without touching them and inspiring them with art.

He led the Arts and Culture workstream in a deadline-pressured effort to create a video that would greet people when they arrived at the Pivot Projects website and invite them to engage more deeply. In one month, the group conceived the work of art, created it, and produced it.

Actually, other members of the group helped with input, but it was really Tom who made it happen. He got loads of help from an artist friend, Dan Wheeler; Max Vladimiroff, a friend of Colin Harrison's who was a composer; and a video editor, Edyta Stepien.

The video told the 13.8-billion-year story of the universe and earth in ninety seconds. It started with a view of the Earth from space being drawn by an artist on a sheet of paper. Holes were punched in the image. Lighter fluid was poured on it. A match was struck. The fire consumed the drawing, but out of the ashes came stunning views of the Milky Way. The musical score expressed the drama and wonder of the narrative. Only three words came on the screen: "Creation. Devastation. Regeneration."

Tom had driven from Chicago to the Rocky Mountains to shoot still and video images of the night sky and the Milky Way—some of which were used in the video. One night in Utah, he borrowed his son's pickup truck and loaded it with photographic equipment. His goal was to drive far up into the Uinta mountains to places where no human lights marred the sky.

As a young person, Tom had suffered from anxiety and depression, and he had worked all his life to live more comfortably in the world. The solitary drive was a test. He was alone, driving on treacherous rocky roads with only his headlights illuminating the path. He didn't know where he was going or if he would be able to find his way back home.

But he made it. He found the perfect spot—a meadow on the top of a mountain with a lake and mountains before him. Above him was the Milky Way. The quiet was intense. "It was amazing to be out in the wilderness alone," Tom recalled later. "I used to have anxiety attacks about loneliness, but the camera gave me permission to stand in a place where I couldn't stand before. It was a transformation. It was spiritual in the deepest sense."

6

The Theory of Everything

When we try to pick out anything by itself, we find it hitched to everything else in the universe.

—JOHN MUIR, NATURALIST

In 2009, Sam Adams, the newly elected mayor of Portland, Oregon, declared in his first state of the city address that he wanted Portland to be the most sustainable city in the world. Over the coming months, he laid out goals for reduction of carbon dioxide emissions and investment in clean energy, and started a program for developing sustainable infrastructure projects, including fifteen miles of bike boulevards. He also launched a massive planning process that brought together more than twenty city and regional agencies with thousands of residents, business leaders, and leaders of nonprofit organizations to help produce the Portland Plan, a set of goals and strategies for making the city more prosperous, educated, healthy, sustainable, and equitable. The planning process took more than two years and included dozens of workshops and hundreds of meetings with community groups.

The mayor's ambitious agenda emerged at a time when IBM was launching its Smarter Planet and Smarter Cities initiatives. IBM made the case that an explosion of digitized data combined with internet connectivity and analytics software made it possible for leaders of government and industry to better understand what was happening in their domains and thus improve the performance of everything, from cities to global supply chains. Further, IBM believed it was possible to model the systems of a city, using data and analytics software, and produce insights that would help leaders make better decisions about the present and the future. The company pitched these ideas to Adams, and he embraced them.

Adams agreed to forge a partnership with IBM to create a digital model of Portland—with the goal of using the insights derived from it to help draft the Portland Plan. It looked like a match made in technology heaven. Portland leaders would get a depth of knowledge about the workings of the city that had not been available previously, and IBM would get a living laboratory for its big ideas. "We were viewing cities as complex systems of systems. This was a chance to put those ideas to the test," said Justin Cook, who headed the engagement for IBM.

IBM's tie with Portland was a tremendously ambitious city-modeling initiative, and it serves as a valuable case study for understanding systems thinking and for exploring the challenges inherent in attempting to model the behavior of complex socioeconomic phenomena. For some of the leaders of Pivot Projects, it was a foundation of knowledge that undergirded their ambitious plans for engaging with cities in the midst of the COVID-19 pandemic. "The experience in Portland teaches that this kind of thing can be done," said Colin Harrison, the Pivot Projects leader who had been the technical visionary for IBM's Smarter Cities initiative. "It also shows that this sort of thing is harder than it looks."

Systems of Systems

Systems thinking (aka systems theory) is arguably one of the more important intellectual breakthroughs of the late twentieth century—and Colin and other leaders of Pivot Projects believed that it was an essential pillar for solving the complex problems of the world and setting society on a path to becoming more sustainable and resilient.

Systems thinking is the notion that our world is made up of a massive number of systems, from microscopic biological processes to urban transit systems, to the vastness of suns and solar systems. Each system, natural or humanmade, is a cohesive cluster of interacting parts. By studying discrete systems in isolation, scientists and social scientists with domain expertise can make sense of them—or at least they believe they can do so. It's when multiple systems interact and influence each other that unpredictability results. Systems thinking is, on one level, a scientific pursuit informed by deep quantitative analysis. On another level, it's a way to help decision makers and leaders understand the interrelatedness of systems in their cities that on first glance might not seem to be related in significant ways.

In the realm of systems thinking, outcomes that are difficult to predict are called emergent behaviors. They sometimes express themselves in the form of unintended consequences, both positive and negative. The idea is that city leaders, city planners, and smart-city technologists should be on the lookout for these consequences when they propose major new initiatives. As an example, when New York City planner Robert Moses built the Cross-Bronx Expressway, he did not expect that it would be a major contributing factor to the rapid decline of the Bronx as a livable city, but it was. The COVID-19 pandemic is another example of emergent behaviors surprising and confounding society. When viewed in retrospect, these consequences seem easily predictable, but at the time they were not.

When the interactions of such systems become highly convoluted, unexpected and often *inexplicable* behaviors occur. Such systems are known as complex adaptive systems, and they are the domain of complexity theory. Dave Snowden, a noted management researcher and complexity expert, likens such systems to a thicket of wild blackberry bushes. He participated in some of Pivot Projects' early meetings. "The concept of entanglement is key in complexity theory," Dave said. "If you pull on one branch you don't know what will happen next." In these situations, it's literally impossible to predict emergent behaviors. This is where resilience comes into play, according to Dave. If your system is resilient, it can withstand surprise disasters. Another tenet of complexity theory is that you have to be able to spot evidence of emergent behaviors early and take quick action to deal with them.

Another key concept in the quiver of complexity theorists is exaptation—as contrasted with adaptation. In biology, exaptation is a radical repurposing under conditions of stress. For example, when dinosaurs evolved, some of them developed colorful feathers, which played a role in sexual display. Then some dinosaurs developed skin flaps on their arms, which were useful in enhancing their performances for potential mates. When animals with feathers and skin flaps fell off a cliff, they glided rather than hitting the ground with a thud. In medical science, drug therapies designed for treating one disease are sometimes found to help treat another. For instance, monoclonal antibodies, which are typically used for treating autoimmune diseases such as rheumatoid arthritis, have been shown in trials to protect people temporarily from the SARS-CoV-2 coronavirus and the COVID-19 disease.[1] In societal systems, exaptation can also be critical. The array of modern collaboration technologies, including Slack and Zoom, were designed to be used by businesses to make money; in the hands of the people at Pivot Projects, they're being exapted to help save the planet. "The essence of exaptation is distributed ideation—trying a lot of experiments and seeing what will work," Dave said. While entanglement sometimes gets us into tough spots, exaptation can help get us out.

Taming Complexity

The leaders of Pivot Projects hoped that they could tame complexity by harnessing collaborative intelligence—with systems thinking and modeling as key ingredients. Systems modeling had been used successfully for understanding and acting on a wide range of complex scenarios, from warfare to corporate strategy—including ambitious urban projects in Portland and other cities. So why not try another approach to modeling for cities and regions in 2020?

All the ingredients for harnessing collaborative intelligence were there. Open innovation (reaching outside an organization for ideas) had emerged as a force for progress with the rise of the internet. The open-source software and open-data movements had shown that by sharing labor and data, we can make the world work better. Now, the tools for real-time collaboration, including videoconferencing and instant messaging, made it possible for people with diverse expertise, points of view, and geographic locations to work together with ease and intensity. Artificial intelligence (AI), which had only recently gone mainstream, offered capabilities that humans don't possess—including the ability to gather and synthesize huge quantities of information, and to see patterns in data that eluded humans. AI promised the ability to learn from the past and predict the future.

One of the salient features of collaborative intelligence is the ability to get things done faster. Collaborative intelligence is also accelerated intelligence.

Colin Harrison envisioned Pivot Projects harnessing these capabilities to become what he called a global observatory. When he worked at the European Organization for Nuclear Research, commonly called CERN, in the 1970s, he watched as physicists from around the world pulled together to design experiments that would challenge or prove the basic laws of physics and related hypotheses. The institution and its devices were shared resources available for use by scientists from more than twenty-five countries. With each discovery, scientists contributed to humanity's shared knowledge of the physical world.

That's what he hoped Pivot Projects could evolve into. In his dreams, the work on the conceptual systems models in the early months would be a down payment on what could become a whole-Earth model, a vast and rich map of the way things work in our world—with many and diverse contributors. "I hope it will develop into an observatory of the entire planet for use by anyone on the planet," Colin said. "Any research team that's interested in using powerful AI, the models, and a vast amount of data could make use of it. They will contribute to our libraries, and we'll use it together."

In physics, scientists are searching for a single, all-encompassing theory that fully explains and links all of the physical aspects of the universe. This is called the theory of everything. It's a mighty big challenge because the two most important theories in physics, general relativity and quantum mechanics, appear to be mutually incompatible at certain scales. Put simply, general relativity operates at the macro level; quantum mechanics operates at the micro level.

I asked Colin if he was comfortable using the "theory of everything" terminology to describe his goals for Pivot Projects—something akin to the quest for understanding the universe in physics. After all, I argued, the group was harnessing systems modeling, open collaboration, and AI to understand how the world works. That was a huge undertaking. No, Harrison said. The label didn't work, not even as a metaphor. His reasoned that, while he and other urbanists study cities to understand how they function, nobody had yet proposed an underlying theory for understanding the mechanisms that govern the lives of cities. "As a result," Colin said. "We don't have a deep understanding of the systems we're trying to improve." I thought about what he said, and then ignored him. The label works for me. It's also a fun thought.

This seems like the place to make an important point about Pivot Projects. While participants expressed a belief that systems thinking, complexity theory, computer modeling, and AI could be useful in improving society, they also had a sense of humor about these things. One day, when Richard James MacCowan was collaborating live on a Zoom call with another participant (Surya Ravi, from India) on a Kumu map, he paused to make a joke: "I came to the conclusion during the Economics workstream meeting yesterday that complexity theory and building models were created by the right wing to confuse the liberals and keep them busy, so they can't get anything done. That way right-wingers can go out and destroy the planet unimpeded."

The Journey of Systems Thinking

Systems thinking and complexity theory started in the second half of the twentieth century. An Austrian biologist, Karl Ludwig von Bertalanffy, developed the general systems theory, a multidisciplinary approach to exploring systems with interacting elements. He focused first on biology and then more broadly on cybernetics—the study of control and communications in living organisms and machines. A few years later, inspired by Darwin's theory of evolution, John H. Holland, a professor of psychology, electrical engineering, and computer science at the University of Michigan, brought his skills to bear on computer science problems related to optimization. He pioneered the field of genetic algorithms, which were software coding structures based on biological principles, and later used biological terms such as *mutation* to explain emergent behaviors in complex computing systems. In the 1970s, British medical scientist James Lovelock introduced the Gaia hypothesis, which states that living organisms interact with each other and their surroundings to form a synergistic system that helps to regulate the conditions that support life on the planet.

Colin got his first taste of systems thinking in 1972, when he was finishing his PhD. At the time, a team of Massachusetts Institute of Technology (MIT) researchers, commissioned by the Club of Rome, published a groundbreaking report, *The Limits to Growth*, in which they warned that the resources on the planet would not be sufficient to support a burgeoning population. The researchers took a systems view of the world. They identified and studied the dominant elements of the world system and the constraints it put on human population and economic activity.

Since *The Limits to Growth* report was published, some of its projections have proven to be accurate, and others are in line with subsequent trends in global population growth and resource consumption. The analysis didn't take into account the potential impacts of climate change because that phenomenon had not yet been identified as an existential threat to humanity.

Colin understood the warning of *The Limits to Growth* viscerally in 1971 when he rode Greyhound buses from New York to Los Angeles to visit Lynn Awtrey, the young woman who would become his wife. For the first time, he saw the vast arid expanses of the American Southwest. "I began to recognize that the planet didn't have infinite resources," he recalled.

It was at CERN that Colin first encountered a highly complicated system, a particle accelerator called the Super Proton Synchrotron (SPS). The device was made of thousands of individual parts, which had to be designed and configured separately and then tuned to work as a complete system. There would be many more experiences with complicated systems in his career. At IBM, for instance, he participated in a project that evaluated the potential for sharing resources among a group of more than 100 petrochemical plants located on an island in Singapore. The idea was that they would recycle and share heat and waste gases. The government loved the idea, but the companies that operated the plants refused to divulge sensitive operational data to one another. As a result, the project went nowhere.

Within the realms of physics, computer science, and engineering, the terms *complicated* and *complex* have different meanings. Complicated things are deterministic, meaning their behaviors can be predicted; complex things are indeterministic, meaning they're very difficult or impossible to predict.

IBM's Smarter Planet initiative is where systems thinking really blossomed—for Colin but also among city planners and government leaders around the world. The company conducted hundreds of engagements with cities, regions, and companies based on the principles of systems thinking and modeling. I mentioned in chapter 2 Colin's leadership of the Urban Systems Collaborative, where he convened meetings of urbanists and systems thinkers from all over. Through that organization, he met Geoffrey West and other grandees from the Santa Fe Institute, which is dedicated to the multidisciplinary study of complex adaptive systems.

West is a theoretical physicist who spent much of his career studying fundamental questions in the field, including the interactions of elementary particles. More recently, he has focused on quantifying the organizational and dynamic aspects of human societies and cities. By assembling multidisciplinary teams of researchers and applying theories of scaling drawn from biology, West and his colleagues have contributed ideas and methods aimed at improving city planning. His most startling finding was that mathematical laws govern some of the important properties of cities, including wealth, crime rates, and the time it takes to travel around, and those factors and many others can be deduced from a single factor: the city's population.

During a workshop on urban science at the Santa Fe Institute, West made a presentation about the scaling laws of cities. In a conversation with Colin and others during the workshop, he jokingly asked whether the physics principle of Hamiltonian mechanics could be applied to cities. The Hamiltonian is a mathematical expression of the various sources and flows of energy in a physical system. These physical systems are deterministic, meaning events that occur with them are determined by previously existing causes. An urban system has many physical elements, but it also has a wide variety of other forces at work, including sociological, financial, cultural, and regulatory factors. They're not deterministic. West's question intrigued Colin. He wondered if it would be possible to quantify and model the basic forces that act within a city, everything from total energy consumption to the impacts of individual choices that people and organizations make every day. It was an exciting thought, but Colin also realized that it was impossible. Not even God could wrap her head around that one.

Finding the Soul of a City

Yet the possibility of quantifying and modeling the basic forces that act within a city set Harrison on a path that has consumed him ever since. He has been on a quest to establish a set of principles about how systems interact within cities, a concept that he calls the soul of the city. He sees the principles defining interactions on four levels—the natural environment, the built environment, the social environment, and the information environment. The goal essentially is to describe an operating system for a city in the way Windows, Mac OS, and Linux are operating systems for personal computers. Computer operating systems control the basic computing functions and act as an interface between computer users and the base technology. In the case of cities, a comparable operating system would be the key to understanding the cities and manipulating them in smart ways.

For Colin, the most compelling element of his vision is the actors in the urban environment: the individuals, private and public enterprises, and civic and community organizations that make choices and act in ways that affect the overall dynamics of the city. He sees those social actors competing with each other at various spatial, temporal, and transactional scales to increase their satisfaction through the choices they make. He believes the soul of the city emerges from their collective drive toward satisfaction. Colin wants to be able to model their motivations, behaviors, and interactions and to discover the role that the flow of information—whether via the internet or through a chat with a friend in a coffee shop—can play in the choices they make and the effects of their choices. One of the greatest values of a city is the density of the flows of information.

Colin is interested in discovering the ways that smart-cities projects, by altering the flows of information, can change the life of the city. "I have been puzzling over this for years," Colin said. "When we do systems models of cities, they're typically the top-down vision of how it seems to be working. But that doesn't explain why it works the way it does. As an engineer, I always want to understand that—to understand the patterns that emerge from all off those individual choices rolling up into something bigger."

In a sense, Colin's quest got its start with IBM's Portland engagement, although he was not directly involved in it. The project showed what was possible and useful in systems thinking and modeling in cities. His quest would advance again, he hoped, with Pivot Projects.

The Portland Project marked a significant milestone in the history of urban planning. The discipline of urban planning goes back more than 100 years. In the mid-twentieth century, the idea emerged that people could use multistakeholder planning processes and quantitative analysis to improve cities. In the United States, a Model Cities Program was launched in the 1960s as an element of the country's Great Society initiative. The program provided funding for new approaches to planning, governing, and redevelopment of slums.

Computer engineer Jay W. Forrester introduced computer modeling to the urban planning world. As a professor at MIT Sloan School of Management, he developed system dynamics, a framework for simulating the interactions between objects in complex adaptive systems, which can be used to address problems in everything from mathematics and physics to economics and urban systems. In the 1960s, Forrester published a book, *Urban Dynamics*, that applied his modeling principles to cities. His theories and techniques were tried in a number of cities but with limited success. Forrester's research team provided the analysis for the Club of Rome's famous *The Limits to Growth* report. To perform their analysis, they used system dynamics modeling techniques.

The Dream of a Better Portland

Over time, urban planners added more quantitative and theoretical tools to their toolkits; however, before Portland, as far as I can tell, nobody had tried to build a comprehensive, data-centric model of a city based on systems theory. It was a joint research project between IBM Research and the city, with no money changing hands. Mayor Adams had a dream of a better, more prosperous, more equitable, and more sustainable Portland. He believed that technology could help turn his dream into a reality. In the technologists and consultants from IBM, he met a group of people who were equally optimistic and ambitious in their goals. At the same time, IBM's Smarter Planet and Smarter Cities initiatives were taking off. The company hoped to build large global businesses based on these ideas and technologies, so it threw resources into the Portland project—and made a lot of promises.

The first task Justin Cook took on when he was assigned to lead the project for IBM was trying to reset expectations. When he arrived on the scene, some city leaders were expecting IBM to produce a policymaking oracle. They figured they would ask questions and get recommendations from the computers—something like the Magic 8 Ball game that was popular in the 1960s. Cook repositioned the technology as a decision-making support system rather than a magical question-answering machine.

Cook and his colleagues used MIT's system dynamics modeling theory and techniques for Portland. (He had learned about the Portland Project when he completed an MBA at MIT in 2009.) The team gathered unprecedented amounts and types of data about a wide variety of systems, including transportation, the economy, housing, education, public safety, healthcare, utilities, and government services. Cook hired an analytics software company, Forio, to help with the modeling. They interviewed several dozen IBM experts on various systems—drilling down on the interdependencies between them. At the same time, they interviewed dozens of Portland government and industry leaders to find out how the city worked. They gathered opinions from citizens at the many public planning forums, and they identified feedback loops within the social systems.

Based on all this input, Forio created a mathematical model of the city. Ultimately, the company displayed the model in the form of a 10,000-node influence diagram. It was similar to a Kumu map. In this case, however, each of the nodes had a mathematical equation underlying it. The equations enabled Forio to perform quantitative analysis. In addition to their models and diagrams, Forio's people used a graphical method for communicating the interrelationships and co-dependencies between the different urban systems. They created hexagon-shaped icons representing systems. Government and community people were able to move the hexagons around to show relationships and illustrate their ideas during brainstorming sessions. A single system could be connected visually to six other systems. "The technology was not a black box answer giver," said Cook. "It helped people see how everything was connected. It was almost a consciousness-raising tool."

In the end, government leaders in Portland found the technology and model to be useful to a certain extent, but they also found it quite frustrating. With the model, department heads could understand better the relationships between their functions, and planners could run what-if scenarios to see how a change, say, in housing development, would affect the current transportation systems. But the technology never really delivered on its promise of providing a comprehensive system-of-systems view of the city. Also, to be valuable on an ongoing basis, the project would have required more data and a continuous flow of data updates. Joe Zehnder, who was then and is now the chief planner for Portland, said, "It was frustrating. In the end, it felt like our leadership had been sold a bill of goods, which they were not able to deliver on. But, still, it influenced our thinking and how we organize and use data." In retrospect, Zehnder wished he had understood better both the science of modeling and the practical application of modeling and simulation to urban situations.

Zehnder said that he enjoyed working with IBM's people, but he felt the company's approach to working with the city was flawed. "City planning isn't an engineering project. It's a political project that's informed by engineering. They didn't get that," he said. Zehnder had the same warning for Pivot Projects. At the time we spoke, Portland had gone through weeks of Black Lives Matter (BLM) protests and had endured waves of violence, injuries, and property destruction. The uprisings had forced city leaders to consider the role of racism in every aspect of living in the city and city governance. Zehnder questioned whether Pivot Projects' approach to problem solving, starting with the models, was broad and flexible enough to address the legacies and structures of racism and income equality in U.S. cities and similar social stresses elsewhere around the world.

Still, Zehnder saw merit in the group's work on systems thinking. He said that those ideas and techniques could be used to help people understand how their cities and communities work and to see their places in them. It's storytelling, really. "They can paint an understandable picture of what it's like to live and do business in the city that people recognize," Zehnder said. "Then they can use the model to peel away some of the false perceptions and assumptions that are causing problems." He believed Pivot Projects could also use systems thinking to examine the potential consequences of policy proposals. For instance, he said, setting carbon reduction goals may seem like a quantitative exercise, like pure engineering. But unless you recognize the impacts of the proposed changes on poor people and minority neighborhoods, you won't be solving the problem in a just and sustainable way.

IBMers who participated in the Portland project and some who observed from afar took lessons from it. Justin Cook, who led the project, acknowledged that the modeling wasn't as useful as he had hoped. "My main takeaway is that the actual process of bringing the agencies and people together, breaking down silos, and talking about interconnecting was super valuable, even if the resulting computer model was less so," he said. Colin Harrison notes that some of the problems encountered by the modelers in Portland have been addressed by more recent advances in technology. The deployment of sensor networks has made more data readily available. Computers and AI can do more of the grunt work of finding data, managing it, and keeping it updated. "One problem was the ambition was ahead of what the technology would easily support," Colin said.

Politics Versus Engineering

The people involved in Pivot Projects had been wrestling from the start with some of the issues Zehnder had raised. On the modeling front, their approach was quite different than the one in Portland. They were making conceptual and relationship models, not quantitative ones. Still, quantitative information and local information would likely be used with the SparkBeyond machine—they just weren't sure how. Overall, at this point, it was unclear how the AI machine worked, how interactions would take place, and what kind of insights would be produced.

At the same time, as a group, they recognized their engineering and science biases. They understood that cities and communities could not be treated like science projects, with the goal of reengineering them to make them work better. Places and their problems had to be viewed and dealt with holistically and collaboratively. That's one of the reasons they had set up workstreams including Faiths; Communities; Economics, Law, and Politics; and Jobs. And it's why they sought participants with diverse backgrounds and points of view. But the core group was made up primarily of engineers and computer scientists, and issues of social justice didn't tend to occupy the prime locations in their frontal lobes. Addressing issues of social justice would remain a challenge.

The Economics workstream was the first to address the role of politics head on. It had started by focusing on economics, developing a theory of change for moving from today's profit- and growth-focused capitalism to a more sustainable economic model. Then the group began to address legal and regulatory structures. But the breakthrough came when they decided to add politics to their agenda and became the Economics, Law, and Politics workstream. That's the way systems thinking works. You may try to handle a topic in relative isolation but then quickly discover that the task is impossible. Everything is connected to everything else.

On August 27, 2020, the Economics, Law, and Politics group launched its foray into the role of politics in making the world more sustainable and resilient. The importance of their explorations had been demonstrated the previous day, when a right-wing teenager armed with an assault rifle shot three people on the streets in Kenosha, Wisconsin, during a BLM protest. New to the group was John Alderdice, a Northern Irish politician and member of the UK's House of Lords who had played a key role in ending twenty years of violence and bringing peace to his country with the 1998 Belfast Agreement. He had helped solve one of the world's seemingly intractable problems, and now he was ready to help take on climate change—another seemingly intractable problem.

On the Zoom call, John introduced himself and his ideas about conflict resolution. He had been trained and had practiced as a psychotherapist but had become interested in why groups of people behave in self-destructive ways, so he moved into politics. In his view, the key to achieving the Irish peace accord was a willingness to include even the most extreme parties in the discussions and also bringing in former U.S. senator George Mitchell, an outsider, as an objective participant. Over time, the parties came to trust Mitchell and asked him to lead the process. Also, confronted with the harm of endless conflict, participants warmed to the idea of compromise. "The lessons are you have to make peace with the extremists, it has to be inclusive, and, also, small interventions can lead to major systemic change," John said.

John had studied complex adaptive systems. He believed that systems thinking would be critical in helping society to pivot to a more sustainable and resilient path forward. A major challenge was that, by and large, people make decisions in the social sphere based on emotions, prejudices, habits, and relationships—not rational thought and data.

"I don't dismiss the computer, but the human dimension is not soft soap," John told the group. "You need it to reach an agreement."

Still, John had bought into Pivot Projects and its process. He saw the COVID-19 pandemic as a moment in history when there was an opportunity to make fundamental changes, the type of opportunity that comes once every few hundred years. He cast back to the day in 1517 when religious reformer Martin Luther nailed to the door of a German church his thesis criticizing practices by leaders of the Catholic church and insisting that individuals, not priests, should be in charge of their relationships with God. That act by Luther was a major turning point toward individualism and democracy. "We moved from kings and bishops deciding things towards ordinary people making their own decisions," John said. "But it hasn't resolved all our problems. We're in search of a new model. I hope Pivot Projects and other groups will enable new models to emerge. We need *something* to emerge."

PROFILE

Anna Panagiotou

Greek archaeologist

When Anna Panagiotou was in high school in her native Greece, she and her parents visited the Palace of Nestor in Pylos, an archaeological site. There, she was captivated by a 3,500-year-old clay bathtub, which had been preserved nearly perfectly. The bathtub was no historical abstraction. She was looking at a commonplace object that had been used by regular human beings thousands of years earlier. She wondered what their lives were like. This experience set her on a path to getting a PhD in Mediterranean archaeology and becoming an archaeologist.

Today, Anna works as a researcher and program manager for the Cynefin Centre, an organization based in Wales dedicated to applying complexity science to public policy and organizational change. She lives and works in Cyprus, where her father's family originated. Anna's focus is on helping organizations understand and navigate complexity. Her primary tool is SenseMaker, which helps organizations and policymakers reach out to regular people and discover what they think about critical issues in organizations, local communities, or the global community.

Those human voices are often absent in decision-making processes, or they're filtered through the lenses of intermediaries. "My greatest hope is that this tool will help policymakers take people seriously in their specificity and complexity," Anna said. "I hope it can be used as a tool for more and more integration of real people at every level of decision making."

Anna joined Pivot Projects on the coattails of the director of the Cynefin Centre, Dave Snowden. Dave had spotted the project on social media. He knew Colin Harrison from IBM, where Dave had been the head of knowledge management. At IBM, he had developed his approach to dealing with complexity,[1] out of which emerged the theoretical framework he named Cynefin (a Welsh word meaning "a sense of place," pronounced kinevin). Dave participated in a few early Pivot Projects meetings, but Anna got more involved. She saw great potential for using SenseMaker in Pivot Projects, and others in the group agreed.

When Dave was at IBM, he noticed that the company caused disruptions in its own operations by introducing rules that were unnecessary or too hard to follow—so employees took short cuts. That caused turbulence, and ultimately the internal management systems began to break down. He saw the similarities between large organizations and other complex systems, such as natural disasters and terrorism, and, applying principles from the natural sciences, storytelling, and complex adaptive systems theory, he developed a framework for evaluating situations. Because it's difficult to choose the best action in complex and confusing situations, Cynefin calls for running multiple experiments simultaneously using different tools and approaches. "This is a sensemaking framework. It helps you make sense of the world so you can react to it," Dave said.

In the midst of the COVID-19 pandemic, the Cynefin Centre was using SenseMaker to help policymakers understand COVID-19 and climate change. Anna was involved in both projects. SenseMaker can be somewhat like a survey because participants are asked to fill out a form either on paper or online. But the folks at Cynefin Centre actually call it an "un-survey" because the respondents' answers are not constrained by the way questions are asked. People are invited to write a short narrative about a theme by describing their own experiences or feelings. Then they interpret their narrative using a series of shapes. They also often provide some standard demographic or contextual information. Anna and her colleagues use a combination of statistical methods and analysis of the narratives to find patterns and outliers. "With SenseMaker," Anna said, "We get a diverse set of viewpoints and we see narratives, not just data points. Sometimes data alone gives you a dangerous sense of security."

The Cynefin Centre's climate project was open to anybody in the world who was interested in participating. The goal was to identify attitudes that, when combined with other research, might lead to collective action by a large number of people aimed at slowing the advance of climate change. One of the COVID-19 projects was aimed at young people (and their parents and teachers), inviting them to write journals about how the coronavirus was affecting their lives.

With Pivot Projects, Anna saw a couple of possibilities for SenseMaker. It could be used internally by the leaders to understand the experiences and points of view of group members. Or it could be employed in engagements with cities or regions to provide a grounded-in-reality sense of what the people who live there want and what they want to change. "People in a community can make sense of what's happening to them. That's empowering," she said.

7

Rethinking Resilience

The plague, like a great fire, if a few houses only are contiguous where it happens, can only burn a few houses; or if it begins in a . . . lone house, can only burn that lone house. . . . But if it begins in a close-built town or city and gets a head, there its fury increases: it rages over the whole place and consumes all it can reach.

—DANIEL DEFOE, *A JOURNAL OF THE PLAGUE YEAR*

On March 11, 2011, an earthquake struck off the coast of the peninsula of Tōhoku, Japan. With a magnitude of 9.1, it was the most powerful quake ever to hit Japan, and the fourth most powerful one ever recorded worldwide. It set off a tsunami that rose more than thirty feet and crashed ashore with tremendous speed and force. All told, the disaster killed more than 15,000 people and destroyed nearly half a million buildings. The quake and tsunami caused meltdowns at three reactors in the Fukushima Daiichi Nuclear Power Plant complex. Residents within twelve miles of the complex were evacuated and hundreds of thousands of them were still displaced years later. Epidemiologists believe that the release of radiation into the atmosphere will likely cause a large number of cases of thyroid cancer within the region over the coming decades. The Great Sendai Earthquake caused about $235 billion in damage, according to the World Bank, making it the costliest natural disaster in history.[1]

Until the COVID-19 pandemic, of course.

When the quake struck, Colin Harrison was sitting in a boarding area at Narita International Airport just outside Tokyo. He had been in Japan attending a technology conference. The terminal shook powerfully for more than three minutes but little damage was done. Peter was shocked when he learned of the disastrous impact on Tōhoku. His flight to the United States made it out that night, but two weeks later he returned to Japan with an assignment from IBM: study how Japan recovered from the disaster. Ultimately, he spent more than three months in Sendai. He learned that Japanese building codes had done a good job of protecting buildings from earthquake damage, but that Japan, perhaps the most resilient society on Earth, had not prepared adequately for the tsunami and its aftermath.

Wreckage from the Great Sendai Earthquake and tsunami

Colin Harrison

There were many lessons about risk and resilience to be learned from the Great Sendai Earthquake, but none was more important than the one that was exposed when ancient rock markers with inscriptions on them were discovered during the cleanup. On the Tōhoku peninsula, parts of the landmass tower more than one hundred feet above the sea, but canyons plunge down to the coast where small villages cling to the shore. When the tsunami hit, seawater fired up the canyons like projectiles out of cannons—wiping out everything in its path. When crews cleaned up the debris, they found stone markers halfway up the canyons. They had been placed hundreds of years ago. They said, essentially, don't build below this spot.[2] "Clearly, a tsunami like that one had struck many years earlier, but, over time, it was forgotten," said Colin. So the greatest lesson to be learned was the simplest and most obvious: don't forget the past.

New Kinds of Disasters

Events and lessons like the Great Sendai Earthquake had prompted the United Nations a decade earlier to create the UN Office of Disaster Risk Reduction (UNDRR) to coordinate risk and disaster planning among member nations. The initiative was formally launched in 1999. In 2005, UNDRR published a document called the "Hyogo Framework for Action" that provided guidelines for prevention, mitigation, preparedness, response, and recovery from natural disasters. The framework was revised and expanded in 2015 as the "Sendai Framework" to include new lessons learned from the Sendai quake and to tie disaster preparedness and response to the UN Sustainable Development Goals (SDGs). The key insight was that, for developing nations to improve life for their citizens, they had to do a better job of dealing with natural disasters, including climate change.

In the science of ecology, resilience is the ability of an ecosystem to respond to disturbances by resisting harm and quickly finding a new equilibrium. Disturbances that are strong or long-lasting can cause ecosystems to begin to break down and eventually to reach a point of no return, where they are unable to recover their equilibrium. We see this now, potentially, in the fragile ecosystem of Greenland, where accelerated melting of its ice sheet has become the largest single contributor to rising sea levels and where scientists now believe massive losses will continue even if there is a decline in surface melt.[3]

Resilience is a sister to sustainability. Resilience focuses on dealing with surprising disruptions that might destabilize a system. Sustainability focuses on long-term stresses, long-term goals, and the strategies to achieve them. So achieving resilience is a key step along the path to sustainability.

Work around ecological resilience has traditionally focused on natural disasters, for example, storms, extreme heat, drought, wildfires, earthquakes, volcanoes, and infectious diseases. But over time, it had become clear that many other factors must be included in the calculus—and COVID-19 sealed the case. The world's abysmal response to COVID-19 showed that we can't improve our resilience unless we also consider a wide array of less tangible factors, including politics and belief systems. During the COVID-19 pandemic, leaders of a number of nations, including the United States and Brazil, publicly denied the seriousness of the disease, were slow to react, reacted inappropriately, and gave incorrect guidance.[4] You might think of these actions and inactions as *un*natural disasters.

They had appalling consequences. Researchers at Columbia University's Mailman School of Public Health estimated that, if social distancing and other control measures had been used in U.S. metropolitan areas one to two weeks earlier, about 36,000 lives might have been saved.[5] That's more than half of the American lives that had been lost during the ten-year span of the Vietnam War. Several months later, in August 2020, when it had become abundantly clear that COVID-19 posed a major threat to health, several hundred thousand people attended a motorcycle rally in the tiny village of Sturgis, South Dakota. Based on news reports and photos, few of them wore masks and observed social distancing guidelines. In the weeks afterward, outbreaks of COVID-19 occurred in South Dakota and the neighboring states of North Dakota and Nebraska. A study estimated that exposures at the rally had led to more than 250,000 infections and cost the U.S. health system more than $12 billion.[6]

This amplification of the disaster's harm was possible in part because of a widespread loss of trust in institutions and social norms. Large numbers of people in the United States and Brazil, for instance, ignored the advice and guidance of medical experts. They had embraced belief systems that included distrust of expertise; denial of science; attachment to conspiracy theories; rejection of articles published by credible journalism organizations; and steadfast adherence to worldviews and opinions expressed by charismatic but dishonest leaders, including Donald Trump and Brazil's Jair Bolsonaro.

Looking ahead from early 2021, it appeared likely that pushback against vaccines might make it very difficult to snuff out COVID-19. Already, there was a strong antivaccine movement among people in developed nations, and opposition to vaccines had been increasing in the developing world. In addition, many people who might have been keen to get COVID-19 vaccinations in an earlier time were expressing doubts about the vaccines being rushed into service by leaders in the United States and Russia.

This belief system phenomenon highlights another major challenge for champions of global resilience and sustainability. As long as populist leaders deny climate science, it will be practically impossible to convince the great masses of people who trust them to acknowledge climate change as a fact much less do anything about it.

COVID-19 also taught other lessons. The spring and summer of 2020 produced a veritable clusterfuck of disasters—to use a technical term. First came the virus, which behaved in unexpected ways; then came the economic shock, which was faster and deeper than national leaders had anticipated. Then politics got in the way as leaders denied the severity of the problem and reacted slowly. Meanwhile, a host of other calamities struck around the world—droughts, fires, hurricanes, and floods. One calamity compounded another.

These entangled events pointed to the need for a wholesale rethinking of societal resilience, according to leaders of Pivot Projects. The traditional approach to planning for resilience was clearly out of date.

That approach focused primarily on one disaster at a time and one place at a time, and mainly on *natural* disasters without fully considering societal influences and amplifiers. It was time to widen the aperture, Pivot Projects leaders said. It was clear that our global challenges are more and more integrated and our responses to them are too fragmented. "It's not just about natural disasters but how we relate to the natural world—the terrible things we do to nature and that nature does to us. We have to look at the whole of it," said Peter Head, who is on the advisory board overseeing the UNDRR's annual global risk assessment reports. He made just such a plea in a video he produced that was published on UNDRR's website in July 2020.[7]

How Connectedness Hurts and Helps

In our modern world, the fundamental nature of contagions has changed. That's primarily because the people and places on Earth are much more tightly connected to one another than in the past and because humans and nature are impinging on each other like never before. We're destroying nature at a rapid clip and it is returning the favor.

Roll back in time to 1665, the year in which the bubonic plague returned to London for the last time. It killed more than one-third of the population; at the tail end of it, the city caught fire and two-thirds of the buildings were destroyed. It was a terrible couple of years for London, but the pain was contained pretty much to a tiny (by today's standards) cluster of people and thatch-roof buildings on the banks of the Thames. Granted, the plague had originally spread from China across Asia and to Europe, and it was a global phenomenon, but it moved gradually along the ancient trade routes. London bounced back. So did other places where populations had been crushed by the plague.

Today, our connectedness is a double-edged sword. We live in a globally integrated society. Because the world's airlines carry about 3 billion passengers in an average year, air travel moves infectious diseases around the world the way the human bloodstream spreads pathogens in the body. The spread of pollution and carbon dioxide is also global. Smokestack ash produced in northern China makes a beeline to the skies of Montana, and chlorofluorocarbons released in California poke holes in the ozone layer above the South Pole. And we all share the ravages of climate change, of course. There's a societal element, too. Because of social media and social networking, lies and propaganda created by politicians, their allies, and state actors spread faster, more broadly, and more effectively than ever before—threatening society with the viruses of ignorance and confusion.

In other ways, though, our connectedness can help promote resilience. The internet and digital communication technologies and cloud computing enabled healthcare leaders to share information more easily about the spread of the coronavirus and to collaborate to create vaccines. Ready access to information from the United Nations and other responsible sources helps national leaders, mayors of cities, and individuals alike size up risks, spot emerging threats, and recover from disasters, both natural and humanmade. Just as connecting more people to one another via communications networks enables the rapid spread of bad information and opinions, it can also be used to spread scientific evidence, information describing solutions to environmental problems, and compelling visions of a more resilient future that would benefit all.

Leaders of Pivot Projects believed that the capabilities they were assembling and putting to work could help make the world more resilient. Collaborative intelligence—that blend of human expertise, computer models, and a powerful artificial intelligence (AI)—was a potent approach to solving the world's most vexing problems. Just like in cities, they believed, systems thinking could be put to work to help leaders and communities solve problems that span regions, countries, and the globe. "COVID has been a resilience test for society all over the world. It has tested almost every aspect of how we live. It has shown where we're not resilient," said Peter Head. "We've seen multiple-hazard risks. We've seen cascades from one disaster to another. These are the kinds of things that systems thinking and modeling are good at helping us to figure out."

Systems Thinking and Resilience

Peter and others involved in Pivot Projects had literally helped write the book on global resilience. Peter and Stephen Passmore, CEO of Resilience Brokers and another Pivot Projects leader, both contribute to UNDRR assessment reports. Stephen, for instance was co-authoring the UNDRR global risk assessment report for 2022, where one of the new themes was the importance of leaders communicating clearly and forcefully about risks. Peter Williams, the former IBMer who led Pivot Projects' Water workstream, co-authored the UNDRR's resilience scorecards for cities and real estate, and helped develop a public health addendum for the cities scorecard.

The two Peters and Stephen are the inheritors of a vector of thought, ecological resilience, that combines study of natural systems with computational analyses. From its start, it has been interwoven with systems theory. Canadian ecologist C. S. (Buzz) Holling, who was instrumental in establishing the modern field of ecology in the 1960s, believed that mathematics could help humans understand the workings of nature so that we could reduce our negative impact on it. For his PhD dissertation, he developed the first math-based theory for predation—working out a formula that calculated the speed with which a predator species in a location would consume its prey based on the density of the population of the prey. That was true, he found, whether you were studying bacteria, hawks, or orcas. He ran his models on an IBM 1130 punch-card computer. (Today's laptop computers are many orders of magnitude more powerful.) One of the surprises he discovered was that this interplay between species didn't always reach an equilibrium. Sometimes it led to the extinction of one of the species. He used simulation modeling combined with policy analysis to develop theories that were designed to be useful in the just-then emerging practice of governmental environmental protection.[8]

Holling was one of the leaders in the study of resilience dynamics, which was the idea that, if you understand the state of a complex system, you could design interventions to help it maintain or regain stability. But he also recognized that complex systems tend to fail at some point,

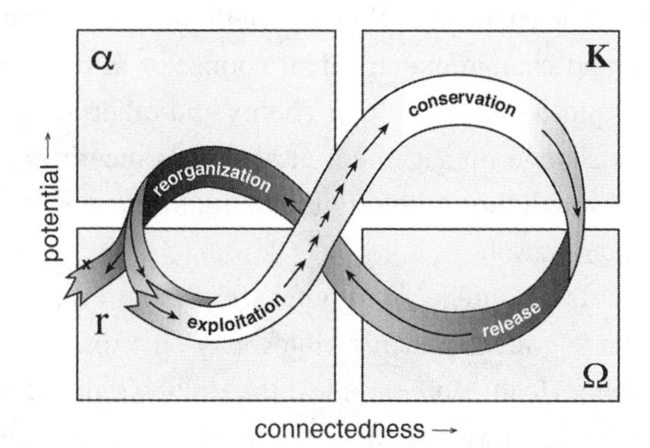

Panarchy model

From *Panarchy*, edited by Lance H. Gunderson and C. S. Holling. Copyright © 2002 Island Press. Reproduced by permission of Island Press, Washington, DC

whether they are arboreal forests in Canada, ocean fisheries, or global financial systems. So he and colleagues designed a model for understanding them, spotting vulnerabilities, and responding to cataclysmic failures. He called it the panarchy model.[9] See the diagram.

Holling preached that all complex systems have life cycles. In the first phase, they gradually build up resources and structure, but they also become more complex and rule bound. He saw that, as systems become mature, they approach invisible thresholds where unpredictable and destructive forces are unleashed, leading potentially to failure. He was an optimist. He believed that during periods of chaotic destruction, ecosystems or societies have the opportunity to reorganize and set off in new and more sustainable directions. He died a few months before the start of the COVID-19 pandemic, but it's easy to see how the spread of COVID-19 and the destruction it wrought fit into the panarchy model. Holling also was a strong believer in the power of bringing together a diverse collection of experts to think about complex problems and to develop innovative solutions.[10]

Notice the similarities between Holling's model and the Two Loop Theory of organizational change developed by Margaret Wheatley and Deborah Frieze of the Berkana Institute. I introduced their approach to managing systemic change in chapter 3. Deborah Rundlett, the Protestant minister and change management counselor in the Faiths workstream, had explored the Two Loop Theory and other life-cycle-based approaches to change management at all-hands meetings, and all of them moved through the phases of growth, maturity, destruction, reorganization, and renewal.

So Holling was the intellectual godfather of Pivot Projects. "I didn't know him, but his ideas and his philosophy—it's totally what we're about," said Peter Head. More significantly, Holling's ideas and the field of ecological resilience that he helped spawn provided the theoretical foundation and the tools of analysis that underpin the Sendai Framework and other approaches to improving ecological resilience.

The United Nations' recognition of the importance of resilience had grown gradually over the decades. It focused first on providing aid to developing nations that suffered from natural disasters, then on prevention and early warning systems. In the 1990s, studies of the El Niño warming phenomenon in the Pacific Ocean made it clear that disasters were global and shared, and had to be approached holistically.[11] The Sendai Framework, which was ratified by the UN General Assembly in 2015, laid out methods and processes for preventing and responding to all kinds of natural or humanmade disasters.[12] Perhaps the most important thing about the Sendai Framework was that it wasn't just another tome fated to gather dust on bureaucrats' bookshelves. It was designed as a guidebook for taking action. At the heart of the project were the scorecards. They provide practical tools that governments can use to assess their disaster resilience, design improvements, and monitor progress. While Holling provided the theoretical foundations for Sendai, people like Peter Williams made it work on the streets, on the seas, in harbors, and on mountaintops around the world.

Putting Systems Thinking to Work

Peter Williams had entered the resilience field from a side door. He got his PhD in political science and worked initially as a political lobbyist and then as a strategy and process consultant for government agencies. IBM purchased Peter's employer in 2002, and a couple of years later, he joined the group that evolved into the Smarter Planet initiative. One of his intellectual guides was Joseph Fiksel, a pioneer of applying resilience and sustainability principles to corporations. Fiksel, who had been a long-time faculty member at The Ohio State University, taught that there's a resilience spectrum, with natural processes and events on one end and human actions on the other. The Sendai earthquake and Hurricane Katrina, which inundated New Orleans in 2005, also helped shape Peter's thinking about resilience. In both cases, planners had failed to spot critical vulnerabilities created by human behavior and practices. He saw that the concept of resilience had to be viewed more holistically, as an intricate dance between nature and humanity. "There's no such thing as a natural disaster," Peter said. "There are events that are rendered disasters because of what people have done—such as building on flood plains, living on fault lines, or cutting down trees—or because of things that we haven't done."

Within the Smarter Planet organization, Peter focused on using systems thinking and information technology to improve water systems. He had the right background for it. At IBM, he had worked on successful projects in California and elsewhere around the world. IBM selected him to be its representative on UNDRR, and that's how he got involved in the Sendai Framework scorecards. He and a couple of coauthors, Michael Nolan and Dale Sands, wrote the Cities scorecard, localizing each of the Ten Essentials—strategies that had been laid out in the Sendai Framework. The authors developed means for measuring the resilience of a city on over 100 dimensions by mapping to the Ten Essentials. Separately, they created a process for action planning.

They also made it clear how the strategies and the systems that they addressed were interrelated. They argued that it was essential for city leaders to develop an overall understanding of the risks and vulnerabilities in the city, to launch a disaster risk reduction planning process, and to bring their new knowledge of risk and resilience to bear on overall city planning and zoning.[13] More than 300 cities around the world have adopted the scorecard. See the diagram for the thumbnail version.

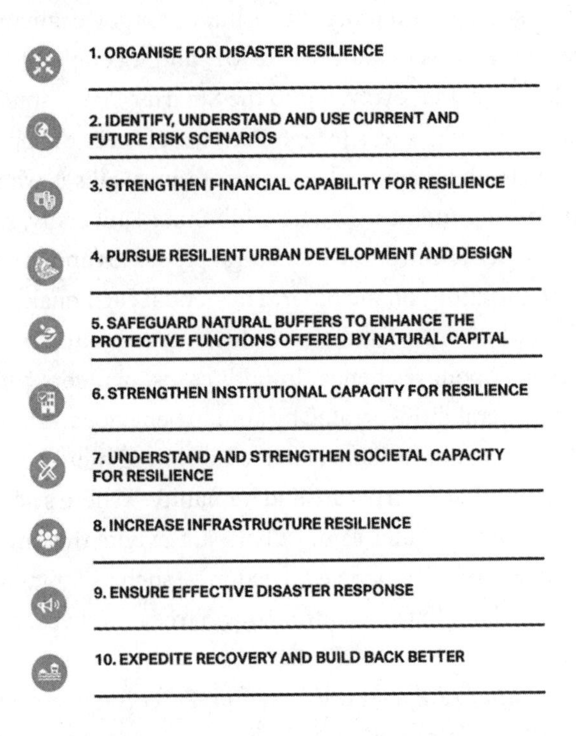

1. ORGANISE FOR DISASTER RESILIENCE
2. IDENTIFY, UNDERSTAND AND USE CURRENT AND FUTURE RISK SCENARIOS
3. STRENGTHEN FINANCIAL CAPABILITY FOR RESILIENCE
4. PURSUE RESILIENT URBAN DEVELOPMENT AND DESIGN
5. SAFEGUARD NATURAL BUFFERS TO ENHANCE THE PROTECTIVE FUNCTIONS OFFERED BY NATURAL CAPITAL
6. STRENGTHEN INSTITUTIONAL CAPACITY FOR RESILIENCE
7. UNDERSTAND AND STRENGTHEN SOCIETAL CAPACITY FOR RESILIENCE
8. INCREASE INFRASTRUCTURE RESILIENCE
9. ENSURE EFFECTIVE DISASTER RESPONSE
10. EXPEDITE RECOVERY AND BUILD BACK BETTER

Disaster resilience scorecard for cities

UN Office for Disaster Risk Reduction

While working on the cities scorecard, Peter noticed that public health hadn't been addressed in the original Ten Essentials (the Sendai strategies), so he and his colleagues volunteered to write a public health addendum, which was published in 2017.[14] That guidance came into play during the COVID-19 pandemic, with additional support and improvement from the World Health Organization. Peter was told that several countries used the addendum to help craft their response to the virus. But in other cases, city and national leaders were caught unprepared. In a number of cities, for instance, leaders had set social distancing guidelines but had failed to prepare their public transit systems to comply with them.

Peter could have predicted this would happen. After his work on the scorecards, he had led a series of city workshops around the world about resilience planning. In one of the sessions, in a New England coastal city, the director of development talked effusively about plans to add tens of thousands of apartment units to the harbor area. Peter recalled that the director of emergency management was taken aback. The director realized that it would be his responsibility to plan for evacuations of all of those people, and nobody had alerted him that this major change was coming to the city's population. The point is that, after years of lectures about the importance of interagency communications and integration, many cities still operated in functional silos. Systems thinking had yet to take hold.

It was possible that the COVID-19 pandemic would change that, at least when it comes to healthcare and resilience. In the midst of the pandemic, the United Nations asked Peter to lead a series of webinars about integrating healthcare with resilience planning. Before the first of them, he expected perhaps 100 people to log in. Instead, there were 1,150 people from 122 countries. Clearly, a large number of city leaders had realized that they could not manage healthcare in isolation because it intersects with everything else. "This has to happen across the board," Peter said. "City agencies can't operate in separate ghettos. Everything they do affects the other systems. They have to work together. You need a high level of coordination and multidisciplinary thinking, and, often, it just isn't there."

A Cluster of Calamities

The insight that everything is interconnected was inescapable in the miserable summer of 2020, at least when it came to disasters. By the end of the summer, there had been about 30 million COVID-19 cases recorded and 1 million deaths. Those totals were still climbing, and vaccines were seemingly months off. The global economy was no longer in freefall, but it had settled into the worst global economic downturn since the 1930s. In rich countries, poor people suffered from malnutrition; in poor countries, the combination of COVID-19 with droughts and other weather disasters caused widespread misery. Global economic integration that had taken a century to develop seemed to be unraveling.

Supply chains broke down, slowing delivery of vital protective equipment. Global pandemic response was fragmented, due in large part to Donald Trump's decision to stop funding or collaborating with the World Health Organization. The world system had been destabilized and seemed to be on the verge of failure.

And that's just the parts related directly to the spread of COVID-19. In the summer of 2020, planet Earth seemed to be suffering from maladies that were entangled with each other and with climate change:

- In July, one-quarter of Bangladesh was submerged due to a confluence of an unusually strong monsoon storm season and sea level rise. Like a number of other low-lying countries, Bangladesh faces a high risk of disruption from sea level rise, but all countries with coastal cities are under threat. An article published in *Nature Communications* warned that, by 2050, sea level rise will push average annual coastal floods higher than land that is now home to 300 million people.[15]
- Droughts and related wildfires destroyed millions of acres of timber and scrub lands in Australia and the west coast and Rocky Mountains in the United States. In Australia, where 2019–2020 was the worst fire season ever, an estimated 3 billion animals were killed or displaced by brushfires.[16] By mid-September, the state of California had already reported its worst-ever fire season, with more than 3 million acres burned.[17] Mass evacuations to crowded shelters put residents at risk of COVID-19 infection, and electrical blackouts related to heatwaves left millions of homes without power.
- Laura, a Category 4 hurricane with wind gusts topping 137 miles per hour swept over Louisiana and Texas, causing an estimated $8 to $12 billion in insured losses to commercial and residential real estate.[18] Scientists believe that warmer ocean temperatures caused in part by climate change will lead to increased intensity of hurricanes, resulting in more damage and deaths.[19]
- In the United States, in the midst of the COVID-19 pandemic, a series of seemingly unwarranted killings of Black people by police set off protests in hundreds of cities. While most were peaceful, some resulted in rioting, looting, and arson. On TV and the internet, the

> whole world was watching. At the same time, COVID-19 disproportionately affected Black and Latino communities due to a variety of factors, including poor access to healthcare. (In developing nations, in some cases, people didn't have access to safe running water to practice handwashing guidelines.) Taken together, police violence and COVID-19 infections sparked an explosion of outrage over the treatment of Black people and the social inequities it fosters.

The breakdown of social norms played out in full public view in Portland, Oregon. Day after day, there were protests over the police killings of Black people. Night after night, small crowds battled with police. Private contractors hired by the U.S. Department of Homeland Security gassed and beat protestors on Portland's streets. On August 30, 2020 a protestor fatally shot a counterprotester. One American man animated by antifascism and racial justice took the life of an American man animated by his Christian faith and the spirit of vigilantism. Days later, the shooter was himself killed by police. He didn't draw a gun, but they killed him the moment they apprehended him. It seemed as if the United States was on the verge of civil war.

Downtown Portland after the unrest of 2020

Marianne Allison

Think about where all of this was taking place. Before COVID-19 and the protests, Portland had been one of the most vibrant cities in the country, and it had long been one of the most progressive. In chapter 6, I described how former mayor Sam Adams had set out to make the city more prosperous, educated, healthy, sustainable, and equitable. That effort had continued with good results. So the worst of the disorder took place in one of the most progressive of America's cities.

Marianne Allison, a friend of mine who lives just outside Portland, drove through the city on the afternoon of September 5, 2020. She shot photos and recorded her observations in a post on Facebook. She noted the boarded-up stores, restaurants, and office buildings, and the fact that there seemed to be more homeless people on the streets. "What struck me most is how empty it is," Allison wrote. "It felt sad in that way." Disaster had struck. The city was in shock. The combination of COVID-19 and civil unrest had hollowed it out at its center.

You had to wonder if the summer of 2020 was a preview of life under climate change: an onslaught of overlapping disasters, natural and unnatural. Did Portland and its people have the resilience to recover and become vibrant and hopeful once again? Did the rest of us? How could we head off an even more dismal future?

Modeling Resilience

The COVID-19 pandemic, combined with the rest of these natural and societal disasters, had helped to create a giant briar patch of trouble on this complex adaptive system we call Earth. The disasters had exposed critical flaws in the defenses of nations and the global community. Poor preparedness, misinformation, poor knowledge sharing, underfunding of governments, poor leadership, and racial and income disparities had made the social systems of the world brittle. And break they did. Scientists and public policy experts said the events proved that governments and societies had to rethink how they prepare for disasters, especially as the effects of climate change accelerate.

The folks at Pivot Projects hoped that computer modeling could play a vital role in helping people understand today's world so we could design a more resilient world of the future. This is the path Peter Head had been on since 2015, when he first proposed creating an enormous computer model representing the critical human systems that figure most prominently in sustainability and resilience. I mentioned his quest in chapter 4. The model was called Resilience.io, and the first pieces were built to help Ghana deal with its water crisis. But after that promising start, funding dried up. Peter had dreamed of combining the Resilience.io platform with full Earth models that were planned at the International Centre for Earth Simulation in Switzerland, which was run by his friend Bob Bishop. The goal of that organization was to help fund and aggregate models of the Earth's natural systems. Peter believed that, by melding capabilities in Resilience.io platform with Bishop's Earth models, they'd be able to help decision makers make better choices in domains ranging from energy and healthcare policy to the design of cities and the prevention of and recovery from disasters.

This melding would harness two domains of the computer modeling world. Resilience.io uses agent-based modeling techniques to simulate the response of the world's systems to perturbations. The agents represent people, things, and events. The agents are expressed in equations embedded in lines of software code that can be adjusted by changing the quantities of people and things and the forces involved in the events being studied. That's how you scale up your analysis from a village to a continent. Bishop's Earth-scale risk models don't use agents. Instead, they collect massive amounts of physical data and then run millions of so-called Monte Carlo statistical simulations to identify the risks inherent in a physical system and the probabilities of a wide range of responses to perturbations.

The availability of vast amounts of cloud computing resources combined with the mainstreaming of AI made these immense modeling tasks more practical and affordable. That's why Peter, ever the optimist, saw so much potential in modeling to help solve the world's problems. AI was especially intriguing to him. Pivot Projects' goal of creating conceptual models of key social and natural systems and exploring potential solutions with SparkBeyond's AI added yet another tool for problem solvers and planners.

Several of the Pivot Projects workstreams addressed resilience. They included Climate Change Risks; COVID Impact and Public Health; Ecology, and Planetary Health; and Water, Floods, and Waste. The leader of the Climate Change Risks workstream was Paul MacLean, who for thirty years had headed an environmental compliance consultancy in Montreal, Canada. He had spent years helping mining companies deal with environmental challenges in particular places, but he saw his group's primary impact being felt at the national government level. He hoped to provide data and arguments that would strengthen the case for large-scale climate action. He expected the Kumu mapping and AI to help clarify thinking and bolster those arguments. "I wake up a little bit unhappy every day because I know what's coming," Paul said. "Working at a city level is important, but how long will it take for that to catch on and have a lot of impact? We don't have a lot of time. I want to focus on having the biggest bang for the buck in a short time."

On the day Paul and I spoke about his aspirations, the skies above major cities in California were blighted with red-orange smoke from the many wildfires that were raging all across the state. It was twilight at noon. People took photos out their windows of a Martian-style hellscape and shared them on Facebook and Instagram. On that day, the whole world could see what is coming. The same thing was happening in Washington and Oregon. Joe Zehnder, the chief planner in Portland, wrote an email to me a day later: "We're watching our state burn around us, which is an event we didn't model for—by the way."

There was a jointly held belief among participants in Pivot Projects in the power of computer models and AI to help deal with climate change and make society more resilient, but there was also a healthy skepticism about the role of machines in problem solving. And that didn't just come from people in the Faiths group. Stephen Passmore, head of Resilience Brokers, said many other factors come into play when community leaders are trying to solve problems. If people don't understand how the machines work and trust their output, it is easy for them to ignore the insights delivered via modeling. "You can have brilliant data and systems modeling, but if it's not received in the right way by policymakers or people in the community, it's just not effective," Stephen said.

To Peter Williams, a key takeaway is the importance of understanding the interactions between the subsystems that together form a large complex system so we can make them more resilient. Scientists and social scientists have a lot of experience in this area. It's not easy, but it's possible. He also highlighted the importance of giving a lot of weight to the human elements in a system—to beliefs, in particular. The best defense against COVID-19 was a mask combined with the willingness to wear it. Many people refused to wear one, seemingly because they didn't trust medical experts or they simply didn't like to be told what to do. Yet to Peter, a critical factor in responding to COVID-19 and in fashioning a more resilient future was the realization that we are all in this together. In a society with nearly 8 billion people, he thought, the balance between personal freedom and personal responsibility is shifting—or needs to shift. "We need to recognize our interconnectedness and our responsibility to one another," Peter said. "We need collective action and collective response based on a clear understanding of what the critical interactions are between each of the systems involved."

PROFILE

Sharmin Shara Mim

Bangladeshi teacher-trainer

In 2015, Sharmin Shara Mim left her family in Kishoreganj, a city in northeastern Bangladesh, to study at a university in the capital city of Dhaka. While at the university, she began to volunteer for an organization that was focused on making the country more resilient. Once, she visited a coastal region that is particularly vulnerable to cyclones and saltwater intrusion. She interviewed people to capture their experiences with storms, flooding, and salinity and, more broadly, to learn about the challenges in their day-to-day lives.

Many of the people she interviewed were women, who, she learned, were paid about half as much as men for doing the same work on fish farms. But she also met a woman who had broken barriers to become a local disaster-risk-reduction manager. The woman went from door to door helping people prepare for the worst. She had organized a group of women who met monthly to discuss their lives. "It was inspiring," Sharmin said. "If there's a problem, they tackle it together."

Sharmin's volunteer work and study visits took her to other parts of the country as well. She saw firsthand how storms, floods, and other environmental problems are affecting everyday people. She could see that climate change posed existential threats to some of the communities she visited. "I'm very concerned," she said. "Eighty percent of our land is flat and near sea level. With sea level rise in the future, these areas will be flooded and a large number of people will be climate migrants. They'll be in desperate condition."

She wanted to help. At the university, she pursued studies in disaster and human security management. After she graduated, she had an internship in a program designed to increase youth engagement in climate change issues. Then she worked as a geography teacher at a middle school in Dhaka. While she was teaching, she learned about Pivot Projects. She saw a post on social media urging people to sign up. She joined several workstreams, including Education and Climate Risks. It was an eye-opening experience. She met people from around the world with a wide variety of points of view. "It has been a fabulous journey for me," Sharmin said. "It has changed my way of thinking. I've learned to be a critical thinker."

Discussions in the Climate Risks group gave her a new take on resilience. Previously, she thought resilience was just about how we recover after disasters, but she came to understand that an even more important element is solving root problems so we can be more resistant to disasters or head them off.

In the Education workstream, she learned to think about practical approaches to learning. She tried to talk about sustainable development to her middle school class., but she found that the students weren't interested. The topic was too abstract. So she changed tactics and asked them to make posters about climate issues.

Working with the students, she realized that the best way to awaken the people of Bangladesh to the dangers of climate change was to teach them when they're young. She and another of the Pivot Projects volunteers, Tu Anh Ha, a Vietnamese teacher, began talking about launching a series of online webinars about environmental risks for university students. Sharmin had been inspired by Greta Thunberg, the Swedish teenager who gained international recognition for protesting and bringing attention to climate change risks. "Greta has shown us how young people can change the world. I want to focus on them," Sharmin said.

In June 2020, Sharmin landed a job as a training facilitator at BRAC Institute of Educational Development, a research outfit in Dhaka. It was a dream job for her. She hopes to study abroad and eventually pursue a career as a university professor, and she saw BRAC as a major step along that path. "It's important to teach about climate change in the university," Sharmin said. "Climate change is happening already. Bangladesh is one of the most vulnerable countries in the world, but few people here are aware of it. We have to change that."

8

Talking to Robots

Of the hundreds of Zoom meetings conducted by Pivot Projects in the spring and summer of 2020, none was more potentially pivotal than the one on September 14. That was the day the leaders of the group met online with Nigel Topping, the UN High-Level Climate Action Champion for COP26, the big climate change conference that was scheduled to take place in Glasgow in November 2021. The event had originally been scheduled for November 2020, but COVID-19 got in the way. Peter Head had been in contact with Topping when Pivot Projects launched, and Topping had recommended that he team up with SparkBeyond for the artificial intelligence (AI) piece. Now, in the wake of Peters's presentation about Pivot Projects to the UK Cabinet Office sustainability group on August 7, Topping had asked for an update.

Until that day, Topping had been something of a mystery for many participants in Pivot Projects. He has been held out by Peter as a potential ally, yet he was notable mainly in his absence. But suddenly, there he was, a postage-stamp-size image on the Zoom screen. He had a bit of a James Bond look, young and vigorous, dressed in a crisp white shirt, with book-laden shelves behind him. Peter, Colin Harrison, and Rick Robinson led the Pivot Projects contingent. Topping was accompanied by Alex Joss, his technology lead.

Topping's role as COP26 champion was to make connections, share knowledge, and generate excitement that might drive climate action. The role had been created at the climate talks in 2015 that culminated in the Paris Agreement, where each participating country set targets for reducing the risks and impacts of climate change.

At the Zoom meeting, Peter updated Topping on Pivot Projects' progress and then the parties zeroed in on ways they could work together. The primary focus was on the UK government's new funding program aimed at helping central and local government agencies use AI to address climate challenges. Prime Minister Boris Johnson's administration had launched the program in early August 2020, and, after Peter's presentation to the Cabinet Office, a government procurement official had invited Peter to participate. The government had set aside £25 million for the program's first year. SparkBeyond had signed up as an approved vendor, with Peter's Resilience Brokers as a subcontractor.

It seemed possible that Pivot Projects and the cities and regions it planned on working with could tap into the funds and use SparkBeyond's AI technology to help analyze sustainability problems and solutions to them. Peter and Topping agreed to look for ways they could work together on behalf of COP26—leveraging SparkBeyond's capabilities.

AI was arguably the brightest and shiniest object in the tech universe in 2020, and many business and government leaders were enthusiastic about it. In a world beset with horrible problems, it seemed like AI could give humans a much-needed helping hand. That was the hope anyway.

The potential of AI had inspired researchers since the first computers emerged in the late 1940s, but the field had repeatedly failed to live up to its promise over the following decades. A turning point came in 2011, when Watson, IBM's question-answering computer system, defeated two past human champions on the TV game show, *Jeopardy!* Watson, presented as a disembodied voice combined with a cool electronic image that played well on TV, brought the wow factor. For decades, humans had fixated on robots as the cultural expression of machine intelligence—both helpful and harmful. Now it was clear that the software inside the machine was where the power resided. It seemed like software robots could transform the world.

The Watson technology was produced by a team within IBM Research led by principal investigator David Ferrucci. The machine (actually a parlor-sized collection of computers on metal racks loaded with the team's DeepQA software) had been named for Thomas J. Watson, the company's founder. Immediately after the TV triumph, IBM launched a massive campaign to bring the power of what it called cognitive computing into practical use across industries and domains, from healthcare and insurance to marketing and national security. The company's aggressive moves seemed to launch a global land rush. Tech companies large and small hurried new AI technologies to market and customers greedily lapped them up.

Colin had been Ferrucci's boss at IBM Research in the 1990s, and he had overseen the early steps in the construction of a technology platform, unstructured information management architecture (UIMA), which was the foundation upon which DeepQA and Watson were built. Colin was no AI expert, and he had been involved in enough big-tech ideas that went nowhere that he was skeptical of grand claims about computer intelligence. Still, he felt the combination of Pivot Projects' expertise and system models with SparkBeyond's technology held great promise. "I think this can be made to work," Colin said. "I hope that we can use SparkBeyond to explore vast numbers of human and natural systems and identify which ones are the best—recognizing that 'best' is a highly subjective term."

Augmenting Intelligence

In the earlier days of AI as a field of study, people thought of it as a potential replacement for human intelligence, similar to the way machines had substituted for manual labor in factories and the way digital tools had reshaped office accounting departments and typing pools. For many researchers, this pursuit was about the grand challenge of seeing if computers could match humans in tests of knowledge or reasoning. The Turing Test, for example, devised by British computer science pioneer Alan Turing, was like a blind tasting test for wine. A human would communicate via computer with another human and a computer. If human number one couldn't tell which of the other two was human, the computer won. It never happened, although Joseph Weizenbaum's ELIZA program, which he developed at the Massachusetts Institute of Technology (MIT) in the mid-1960s, fooled some people. Other scientists focused on capturing the knowledge of experts, storing it, and doling it out as needed to decision makers in organizations. Who needs an expert when you can have an expert computer? These so-called expert systems flopped. Today, some computer scientists and futurists predict a time, which they call The Singularity, where machines will overtake humans in the brains category and master general human intelligence. This time is only a matter of years away, they believe.

In my days at IBM, I helped coin the term *cognitive computing*. My colleagues and I wanted to avoid using the term *artificial intelligence* because it was so closely associated with specific fields of academic study, including machine learning, natural language process, neural networks, and the like. Cognitive computing tapped a broader set of technologies, and our scientists and programmers were doing real-world stuff, not theoretical. Also, we didn't want to be associated with Hal, the computer on board a spacecraft in Stanley Kubrick's 1968 science fiction blockbuster, *2001: A Space Odyssey*; Hal decided that the human astronauts presented a threat to the mission and ignored their commands. Computer intelligence would not replace human intelligence, we asserted. They were complimentary. Humans and machines would collaborate to accomplish things that neither could do as well on their own.[1] A few decades (or years) from now, this assessment might well look quaint. But at the time, and still today, it seems like the most constructive spin to put on the fast-evolving relationship between smart machines and humans.

This was the view embraced by the leaders of Pivot Projects and SparkBeyond. The terminology had evolved by then. They called it augmented intelligence rather than cognitive computing. Colin had even written a book chapter about it.[2] "I see AI as systems that will often operate independently, whereas I see augmented intelligence as a modern equivalent of a machine tool. We do not wish to give them agency and thereby lose our hold on the sense of values or morality we associate with humans. We see them as extensions of human users rather than autonomous agents," Colin said. Shay Hershkovitz, the vice president at

SparkBeyond who was the primary liaison with Pivot Projects, put it this way: "Human intelligence is amazing, but we have cognitive limitations and biases. AI is amazing because it can quickly sift through and analyze large quantities of data, but the ability to reason is limited. It relies on statistical inferences. But if you combine the two, you get a good result. You allow humans to do what they do best and machines to do what they do best."

It is intriguing to envision a computer as an intelligent assistant or adviser. Computers are increasingly being asked to play these roles. But given the limits of today's technology and probably tomorrow's, even the best AI machines should not be viewed as infallible soothsayers. The scientific quest to create machines that possess general human intelligence is still akin to medieval knights fruitlessly searching for the Holy Grail. Rather, the state of the art in AI-based decision support systems rests in systems that are designed to interact with humans—responding to natural language queries with recommendations, evidence, and suggestions for new avenues of inquiry. Think of it as guided discovery.

One of the more mature examples of this type of application is Watson for Oncology, a recommendation engine developed by IBM and Memorial Sloan Kettering Cancer Center for use by doctors treating cancer patients. First introduced in 2014, the system has been deployed in more than a dozen countries worldwide and has shown strong performance for treatment of lung, breast, and ovarian cancer but weaker results for other cancers.[3] The system isn't a replacement for a highly trained oncologist, but it can help doctors identify the most appropriate treatments for individuals, and it can be especially useful in places where oncologists aren't readily available. The mixed bag of results achieved by Watson for Oncology shows just how challenging it is to develop AI machines that think as well as humans.

Colin, Peter, and others within Pivot Projects were well aware of the shortcomings of AI, but they were also excited by the possibilities. SparkBeyond's paying customers had used the system successfully for everything from selecting M&A targets to finding new uses for their core innovations. Also, Pivot Projects would be using the machine to explore ways to improve sustainability over a period of months or years, not making immediate life-and-death medical decisions. There was more moral wiggle room. "I believe it can be a game changer, but it will only be as good as we make it," Peter said when Pivot Projects volunteers began engaging in earnest with the SparkBeyond machine. "We're setting up a learning loop. If we start down the road and it's not working, we'll adjust. We'll retrain it. We'll teach it to do better."

SparkBeyond's Story

As a journalist who writes extensively about the tech industry, I'm attuned to an interesting social pattern: when interviewing Israeli tech-company executives, I frequently discover that they learned their trade while completing their national service in an intelligence branch of the Israel Defense Force. Typically, they aren't permitted to tell you much about what they did when they were in uniform. And so it was with Sergey (Sagie) Davidovich, cofounder and CEO of SparkBeyond. Sagie served in the Israeli air force and actually launched a tech start-up there, raising funds and building neural networking software for diagnosing problems in computer systems. The drives to invent and build have long cohabitated in his brain.

Sagie's roots as a technologist go back to his childhood. He was born in Ukraine, not far from the Chernobyl nuclear power plant. When he was three years old, the plant melted down, spewing radiation and triggering mass evacuations. His earliest memory was that he couldn't eat strawberries anymore. The family left, eventually settling in Haifa, where Sagie's parents bought him a personal computer when he was seven years old. He had no friends. He could not speak Hebrew at first, so he lost himself in technology. He loved to write computer games, and before too long he was selling them. Once, at a weekend programming contest, he wrote a software program for translating typed messages into music for communicating with imaginary beings from another planet.

Intermixed with his military service, Sagie had studied electrical engineering and bioinformatics. In his twenties, he launched a couple of tech start-ups and held leadership positions in others. Then, in 2013, he started talking with Ron Karidi, a software company executive who had previously run Microsoft's lab in Israel, about launching an AI-based start-up. The idea was as grand as they come: to solve planetary-scale challenges by combining humanity's knowledge with AI. "I'm the dreamer and he's the pragmatist," Sagie said. "I think of crazy ambitious ideas and he's good at critical thinking. We complement each other." Later that year, SparkBeyond was born.

They began by developing a Web crawler that software developers could use to find pieces of open-source software code that they could add to their own projects rather than starting from scratch. What Google does for text, their first machine did for code. That technology evolved into what Sagie calls a problem-solving engine. In its mature form, the technology would help organizations formulate or test new hypotheses, explore complex fields of knowledge, discover novel solutions to problems, and the like. It would overcome cognitive biases and data-processing bottlenecks that limit human problem solving. In short, it was an ideation machine. "Other AI technologies help you answer questions. We do that and more. We help you ask the right questions," Sagie said.

The system took several years to develop fully. It wasn't until 2015 that SparkBeyond declared the technology to be enterprise ready, and it's still evolving rapidly today. Along the way, the company won a series of high-performance computing contests and forged partnerships with Microsoft and several consulting firms, including McKinsey & Co. The company has customers in a wide variety of industries, including insurance, banking, and healthcare.

The SparkBeyond platform employs a number of AI techniques. The software discovers and optimizes information for exploration, discovery, and analysis. A key element of the platform is the knowledge graph, which describes real-world entities and the relationships between them using ontologies, which play the role that grammar does in human language.[4] Each of the nodes in a knowledge graph represents an entity of interest. In SparkBeyond's knowledge graph, the nodes are categorized as resources, processes, or attributes. Among other approaches, Kumu can be used for visualizing elements of its knowledge graph.

SparkBeyond gathers information from a wide variety of sources, including Wikipedia, the online open-source encyclopedia; other data that is freely available on the web because it uses the Linked Data format devised by Internet pioneer Tim Berners-Lee; and a variety of other data sources, including patent records and scientific article repositories. In addition, the machine can connect to any number of private or government data sources.

After users of the SparkBeyond technology define the datasets they want to investigate and apply their ontologies, the system uses machine learning and other AI techniques to reason about the data, derive insights, and help humans explore hypotheses. The system actually has two applications. One, called Research Studio, is used for mining knowledge from the web, identifying the principles underlying patents, and analyzing trends in business and society. The other application, Discovery Platform, is used for creating and evaluating hypotheses using highly structured data. The applications can be used separately or combined. The hypothesis engine is designed to be used by hard-core data scientists for subjecting hypotheses to math-based analysis, but Research Studio can also be used to develop and test hypotheses. Because of the way the system is tuned, it typically delivers highly relevant results to queries in anywhere from less than a second to a few seconds.

Some of the techniques SparkBeyond uses are cutting edge. The Discovery Platform, for instance, uses a type of genetic algorithm to produce highly relevant search results. In chapter 6, I told you about John Henry Holland, the professor at the University of Michigan who applied complexity theory to computer science and who pioneered algorithms modeled on the Theory of Evolution. The field has evolved considerably since Holland did his initial work in the 1970s. SparkBeyond's genetic algorithms convert ideas and hypotheses into computer code. That makes it possible for the system to test hypotheses quickly by evaluating large quantities of evidence, both pro and con. If a hypothesis performs well in this first phase, the system launches a second phase of analysis where it makes thousands of copies of the hypothesis structure and introduces random changes into them, then tests the results against a predefined standard of fitness. The system repeats this process multiple times, learning from the results and continually optimizing the hypothesis until it is fully tested and proven or disproven. This technique not only enables users of the system to stress-test their hypotheses; it also enables them to discover novel solutions to the problems they're seeking to solve. Another feature that's beneficial both to SparkBeyond and its customers is that, because of the system's machine learning technologies, it improves with use.

One of the most challenging aspects of using the system is the fact that human language and computer language function quite differently. Human language is complex and imprecise. Words can have multiple meanings or shades of meanings. Context is critically important to understanding what a person means when she or he says or writes something. In comparison, computer language is formulaic and precise. The machine doesn't understand context unless it is given a lot of lessons. With humans, the more words we use, the better we understand each other. With machines, it's the opposite. "We humans thrive on ambiguity. We have context and common sense to help us. Machines collapse when confronted with ambiguity," Sagie said.

The field of natural language processing has the goal of enabling humans and machines to communicate effectively, and a tremendous amount of progress has been made over the past several decades. But that interplay between human and machine is even more challenging when humans are communicating with computers through models. The best models of systems are those that use the most precise words, preferably nouns and verbs. Yet we humans feel like we haven't explained things properly unless we serve up a word salad with lots of verbs, adverbs, and adjectives. This would be one of the chief difficulties that Pivot Projects volunteers faced as they melded their models with the SparkBeyond machine.

In most other ways, fortunately, the two organizations meshed nearly perfectly. From its start, SparkBeyond has been driven by a social impact mission. In 2019, the company formalized its mission statement: "to harness humanity's collective intelligence to solve the world's most complex challenges." It was the mission match that attracted Sagie and Shay to work with Pivot Projects. In January 2020, SparkBeyond had hired Shay to run a new initiative aimed at using the company's technologies to help address climate challenges. When the opportunity to forge a partnership with Pivot Projects came along, they jumped on it. Pivot Projects had the goal of helping to put society on a more sustainable path and had a diverse group of volunteers dedicated to making it happen. SparkBeyond had technology that seemed to be custom-designed for exactly that kind of exploration.

During the COVID-19 pandemic, it had become abundantly clear that society is not good at solving complex systemic problems. In the early months of the pandemic, governments spent trillions of dollars to fight COVID-19 and soften its economic impact. The best minds in epidemiology and public health had failed to stop its spread, and experts in economics and crisis management had not been able to repair its damage. On October 2, 2020, Donald Trump announced he had tested positive for the disease, and, later that day, he was rushed to the hospital in a helicopter. Nearly a year after COVID-19 had begun its spread, the future of the disease and of society were still highly uncertain. We humans needed help. Could AI provide it by filling in the blanks in our capabilities, augmenting our knowledge and thinking? Pivot Projects wasn't designed to solve the COVID-19 pandemic, but it addressed an even larger scale of complexity and uncertainty. The hope was that this combination of human and machine intelligence would help with the longer-term recovery from COVID-19 and with setting a more sustainable path forward for society.

Talking to Robots

A Zoom meeting that took place on October 5, 2020, was a bit like an AI 101 seminar in college. Ian Abbott-Donnelly, Pivot Projects' Kumu and AI maven, had been learning about SparkBeyond's AI machine for months, but just a couple of weeks earlier, Shay had tossed him the keys to the car and invited him to take it for a spin. Ian did so, and he was delighted that he and a small group of young Pivot Projects volunteers would finally be able to work directly with the AI. This session was supposed to help them answer questions that had arisen when they experimented with the system. It's important to note that while SparkBeyond's Research Studio, the primary application they were using initially, possessed a wide range of technical capabilities, it did not yet have a fully developed user interface or an automated technique for ingesting Pivot Projects' system models. A lot of the work for paying clients using the machine was done by SparkBeyond's deeply experienced technologists. Ian and his young colleagues were to be design partners for the product: the people who experiment with it in its rough form and help the designers improve the look, feel, and functionality. They would help SparkBeyond create a tool that practically any well-educated human could use.

Ian had collected a half-dozen key questions that he and other Pivot Projects volunteers had raised as they experimented with the system. With Ian on the call were James Green and Andre Hamm, who had been his principal helpers on the Kumu mapping project, and Bryony Bowman, a PhD candidate at the University of Birmingham in the United Kingdom who was studying sustainability and resource recovery. SparkBeyond was represented by Ron Oren, Den Wainshtein, and Shay Hershkovitz. Ron was the product lead. In addition to his day job at SparkBeyond, he had just days earlier launched a start-up on the side—ImagineAI. He was developing an algorithm that learns professional photo editors' workstyles and automates processes for them.

As with Kumu, Ian's approach to the SparkBeyond machine was to learn by doing. He and his team had been exploring the system; pressing digital buttons and seeing what happened; and taking notes when they came upon something they didn't understand, which was relatively often. SparkBeyond's user interface was far from intuitive. Ian prefaced the interaction with a status report of sorts: "We're at a stage where we have our eyes wide open but we're not understanding what we're seeing all the time."

They peppered the SparkBeyond team with their questions. The Israelis provided quick and direct answers. It was a bit like watching a Ping-Pong match between friendly players who didn't care who won. Clarity arrived in digestible chunks. The pace slowed a bit when Ian asked for help in posing questions and in filtering to get the most useful results. Shay's advice: "I try to frame it as a question at the top, then make a list of subquestions. By answering them, you answer the overall question." He also urged them to keep their questions as simple as possible, and he gave some detailed advice on setting filters that would focus the results of searches. The meeting ended abruptly when the SparkBeyond team ran out of time and had to rush off to other meetings. Ian and his colleagues stayed on to discuss techniques for loading the first Kumu models into the SparkBeyond machine. A few hours later, looking back, Ian said he was beginning to get a better idea of what a workshop involving Pivot Projects and city leaders would be like. "This is much more than search. We're actually computing with the ideas," he said. "Our challenge is to ask good questions. SparkBeyond's challenge is to make the big answers understandable."

Developing the Water Pivot

Three days later, Ian loaded one of the Kumu models into the SparkBeyond machine. SparkBeyond's engineers were working on an automated technique for loading Kumu models, but it wasn't quite ready yet, so Ian had to load the model manually. His quest was to discover a more fair and effective approach to pricing water. The Water workstream had wrestled with a moral conundrum: they believed that access to clean water should be a right, yet they recognized that water is a limited resource in many locations and thus it should not be wasted. In discussions, they had explored the possibilities for water systems operators offering a basic per-household ration of water for free and then charging for higher-volume use of water. This kind of approach had been used only rarely by water authorities around the world, but it seemed promising. Could this pricing model enable society to provide everyone with the water they need while ensuring the ready supply of this vital resource? That was the question that Ian began exploring on his own one evening.

By asking simple questions, Ian found an example of a water district in Belgium using this kind of pricing approach. He also found a trove of news articles and scientific papers examining water pricing and drought. He could see how the tool could be used in real time by a group of experts to answer the rest of his questions. "The one experiment in Belgium could be a weak signal of a big change that needs to happen everywhere," Ian said. "The topic is worth a lot more exploring."

The combination of human and machine intelligence is what makes these kinds of explorations so potentially fruitful. It starts with building systems models. By creating and uploading Kumu models, Pivot Projects' humans make it clear to the machine what is most interesting to them, the most important concepts operating within a system, how they see systems operating, and the important relationships they see within systems or between related systems. Sagie calls system models the scaffolding for augmented intelligence. Once the models are loaded into the AI platform and people start asking questions, the machine reaches out into its universe of information sources to build richer models. Fed with additional information, the machine will discover new entities that should be considered and new relationships between entities. It is designed to correct for misleading biases or misconceptions introduced by the humans. Through this process, combined with the ongoing interactions of humans and the machine, the AI platform gradually builds a digital twin of the real world that can be used to simulate and explore it. "This is a massive undertaking," Sagie said. "We want to get as close as possible to simulating the dynamics of the world."

Over a number of weeks, Ian worked with Peter Williams, head of the Water workstream, and other Pivot Projects volunteers to use SparkBeyond's machine to develop the water-pricing hypothesis. They called their project a pivot. They hoped to be able to use this work as a demonstration of the capabilities of Pivot Projects and SparkBeyond that they could present in initial meetings in communities with which they hoped to engage. This was mostly Ian interacting directly with the AI and discussing with others the results he found rather than the real-time group collaboration sessions they looked forward to in the future.

Ian was no water expert going in. He was a generalist with a good grasp of the scientific method and computer-aided research techniques. With the help of the AI and his more-expert colleagues, however, he was able to learn a lot about the topic in a very short time. "This is where the real value of the AI comes in," Ian said. "You can go from a position of complete ignorance to pretty broad knowledge in a couple of weeks."

In retirement, Ian was a bit of a Renaissance man. He was a member of a road biking club and he was a runner. He took painting classes; enjoyed time with his family; did some research projects with a group at nearby Cambridge University; and, of course, was a near-constant presence in Pivot Projects. He tended to do his work on the Water Pivot in the evenings. Picture a guy working alone in a room surrounded by books, paints, paintings, and electronic gadgets. He spent hours there over a couple of weeks exploring the nooks and crannies of SparkBeyond's AI.

In SparkBeyond's user interface, you type questions just like with a Google search, but you can also choose among a number of factors that will help you filter results and focus your explorations. Those factors include the ability to describe key concepts and to specify the kinds of relationships between them that you want to examine. In addition, the interface allows you to do specific kinds of searches. One of them, Pathfinding, enables the user to discover indirect connections between one concept and another in order to help validate or refute hypotheses. Another, Co-occurrences, enables you to analyze the intersections between concepts across a wide variety of sources, including patents, scientific research publications, news, grants, medical clinical trials, and more.

The first thing that Ian discovered was that, while water management might seem to be simple enough on the surface, it was what he called a "complex wicked problem." Once you start taking a systems view, the tangled web of interrelationships between concepts becomes apparent. Dozens of factors are at play when you examine the relationships between water pricing, scarcity, pollution, and carbon impact.

Over time, Ian used the tool to pull together a view of the standard business model for water management—which is unsustainable. Typically, residential and commercial customers pay low prices for water, which makes it difficult for authorities to fund long-term infrastructure improvements. A lot of money is wasted because of leakage and unnecessary pumping of water from point A to point B. Meanwhile, people pay

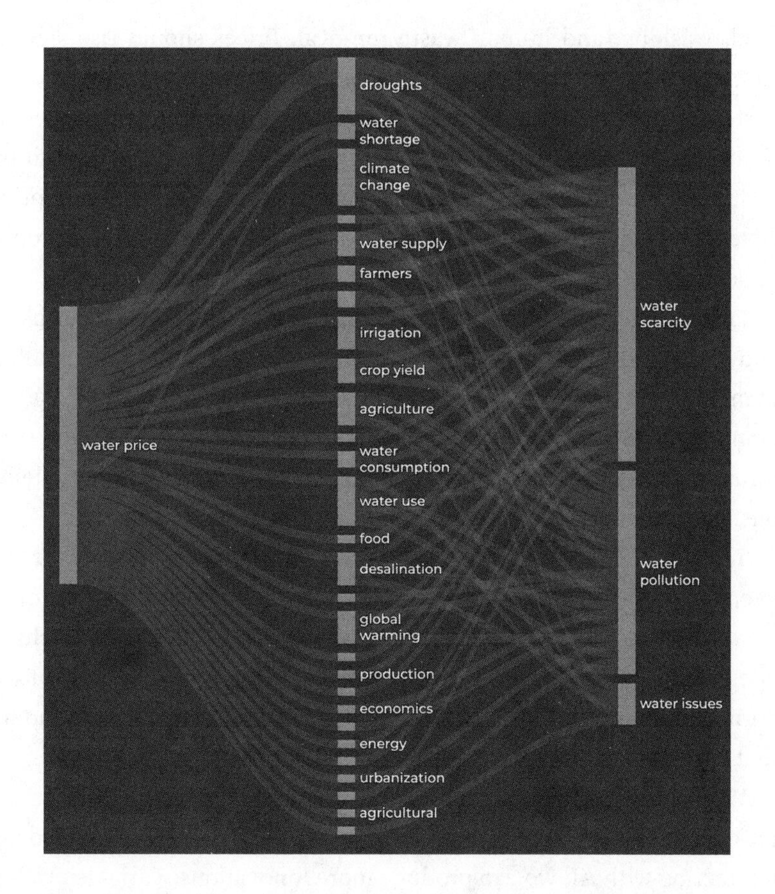

View of water-pricing relationships in SparkBeyond's AI software program

high prices for clean water in plastic bottles, the production of which burns a lot of energy and produces a lot of waste. In both cases, the true environmental cost of water consumption is not accounted for, resulting in waste and pollution of this essential but finite natural resource.

Then, Ian and his colleagues produced the outline of the Water Pivot. Water pricing must be done with a systems view, they argued. All information about water supply, use, and associated costs should be gathered and made transparent. All of the stakeholders in a particular catchment area should have a say in how the water there is priced. A pricing structure should be set that ensures that every person living in the catchment has the water required for the basic human needs of drinking, preparing food, washing, and human waste removal. Prices should rise significantly and on a steadily rising scale for additional water usage and for customers that degrade the quality of the water they use. The full environmental costs of supplying, managing, and treating water should be included in pricing. Ian and his colleagues believed that these core principles would lay the groundwork for an approach to water use that would be sustainable, regenerative, and fair.

Ian, working with the AI and his colleagues, had evaluated a problem and developed a potential solution that was different and more detailed than any of the humans involved had expected going in. It was a good demonstration of the power of collaborative intelligence. The next step for Ian would be to turn his explorations into a crisp, step-by-step demonstration that he could show to potential engagement partners.

The Water Pivot exploration had been a bit of a mind blower for the people who participated in it. They saw that this kind of research could turn up all sorts of novel solutions to problems. Because the machine doesn't have the same biases that we have, it can find insights that we normally might not be exposed to. It doesn't have the cognitive blinders that humans wear.

We might even find potential solutions to problems we aren't even looking at. Serendipity is an important factor in innovation, and by interacting with AI, we can produce more innovations. Variable pricing schemes, for instance, could be used in all sorts of situations to make society more resilient and sustainable. Here's an example: today, in many cities, pedestrians and bicyclists are struck by motor vehicles all too frequently. One solution is to install automated traffic sensors that detect stoplight, stop sign, and speeding violations. But there's concern about fairness. Wouldn't the system have the greatest impact on poor people, who can't afford to pay large fines? To deal with that, the traffic authority could design a variable pricing system that levies fines based on the driver's ability to pay. In this way, driving behavior could be modified at scale, the lives of pedestrians and bicyclists could be saved, and it could be done without oppressing people who already have a difficult time in life. In addition, while we're using the AI machine to examine variable pricing schemes for water, we might find another solution that helps our community shift from cars to walking and bike riding, which are more environmentally sustainable.

Over Ian's decades as a technologist, he had used all manner of computer-aided research tools and search engines. This research tool seemed to him to represent a major step forward. "Search finds you *things*," Ian said. "The AI points out the *relationships* between things, and relationships are much more valuable than things when you're taking a systemic view. We're bringing together data, knowledge, people, and the AI to have this rapid, systemic, collective intelligence."

A Practical Tool for Exploration

Ian was an experienced computer scientist and model builder. Without too much trouble, he had SparkBeyond's technology working for him. But would the AI work for the rest of us mere mortals? Ian had recruited a group of young people to become Pivot Projects' AI guinea pigs. He and SparkBeyond's technologists provided lessons and tutorials, but they also let them play with the computer on their own to learn by doing.

One of the learners was Chris Medary, a graduate student in environmental management at Western Colorado University. He had joined the Climate Risks workstream, among others. Even before he arrived at Western Colorado, he had been drawn to the lessons we can learn from indigenous cultures. In Colorado, in what had formerly been Ute country, he read and thought more deeply about these issues.

Chris was particularly interested in acequia culture, an agricultural tradition that originated in central Asia 10,000 years ago and flourished in the U.S. Southwest after the arrival of the Spanish in the sixteenth century. The Native Americans and the Spanish invaders adapted acequia to the arid Southwest, combining gravity irrigation and land terracing with crop rotation techniques to produce maize, squash, legumes, and peppers in abundance. There's also a social aspect to the culture. Acequia communities manage resources like water and land and labor together for mutual benefit. Chris decided to use SparkBeyond's AI to explore modern uses of acequia to achieve more sustainable agriculture.

He was a little intimidated at first. Chris had no idea how the damned thing worked, but then he discovered that the machinery of AI was irrelevant to him as long as the system helped him do what he wanted to do. He also liked the fact that SparkBeyond wasn't an autonomous system. Explorations are done jointly by humans and the machine. He found that the interface was easy to learn and to use. He launched explorations by typing in keywords, starting with *acequia*, of course. His searches led him to a host of articles from a variety of realms, including indigenous agriculture, communal culture, history, and modern examples of acequia in practice. He figured he saved hours of time by using the SparkBeyond platform and discovered sources he might not have found otherwise. The key insight that he came away with was that, just as the Spanish and the indigenous people in the Southwest had created a hybrid approach to agriculture, modern farmers could combine acequia principles and practices with modern farming techniques and technologies to produce crops both efficiently and sustainably.

"There's always a debate in the environmental field: do we just return to the land or embrace the techno future," Chris said. "This is the perfect example of doing both. You find the parts of each that seem to work the best. It's a great lesson for people in the sustainability field." Based on his research, he planned on writing a case study of the potential for harvesting ideas from the acequia tradition in a report being produced by the Climate Risks workstream.

AI and Ethics

Whenever AI comes up in conversation, ethical issues often emerge. That's because humans are justifiably concerned about the impact of AI on us and our world. AI is tremendously powerful already, and this is just the start. It will be what we make of it, and, we are a motley crew of individuals and organizations. Many of us are inept or careless, at least occasionally. Others are corrupt; criminal; oppressive; destructive to the natural environment; or, in a few cases, intent on world domination. AI can be a tool for good or evil, just like atomic energy, and thus we have to guide, monitor, and govern its use. The topic is huge, and I won't attempt to grapple with all of its dimensions here, but one thing is abundantly clear to me from my years of writing about AI: every organization that develops or uses AI should follow a clear set of guidelines. Granted, the world of AI is constantly changing, so the issues that we have to grapple with will change, but organizations need to agree on core principles, follow them, and adjust them as necessary.

Many of the organizations that develop or use AI heavily, including tech companies and major industrial companies, have adopted principles for using AI. SparkBeyond is among them. In 2019, the company's leaders worked with employees to develop a document they called their moral compass. It included guidelines for who they would work with and how. Potential customers are slotted into color-coded zones based on their purpose and behavior. Green is for organizations with which the company actively seeks to engage, especially those that address the UN's Sustainable Development Goals (SDGs). Blue is for outfits that aren't especially progressive but do little harm to the environment or humanity. Orange is for organizations that engage in activities that SparkBeyond doesn't like, such as fossil fuel companies. It will work with orange organizations if the project helps drive a shift to alternative energy sources. Gray, another borderline zone, includes national defense and intelligence agencies. SparkBeyond might work with such an agency on a project that it deems to be constructive, such as public safety, but it won't license its technology to them for other uses. Last is the red zone, which includes the weapons, tobacco, and gambling industries.

The company's ethical guidelines for deploying AI are much less detailed. They're situational and depend on the judgment of the company's leaders:

- Fairness, privacy, accountability, transparency: We use our collective judgment to act according to, and to ensure a fair balance between, the ethical dimensions of applied AI. We help our partners to address these dimensions and the trade-offs.

- Thought leadership: We promote thought leadership in raising matters of ethics in our day-to-day, and we proactively engage as a team and with our partners to resolve issues.

In my view, it would be better for them—and every player in AI—to spell out in greater detail their ethical guidelines so that they are clear for employees and customers alike. A number of companies have done so, and some have acted on their beliefs, even though it cost them financially. IBM, for instance, abruptly got out of the business of making facial recognition software in the wake of the Black Lives Matter (BLM) movement in the United States over concerns that the technology might be used in a way that would enable racial and gender bias. Microsoft also set a high ethical bar. The tech giant established an oversight committee and a set of working groups to keep on top of the shifting landscape of AI ethics. Its Office of Responsible AI puts its principles into practice. One of the soundest and most explicit statements of AI ethics I have found is from the U.S. Department of Defense (DOD), adopted in 2020:

- Responsible: DOD personnel will exercise appropriate levels of judgment and care while remaining responsible for the development, deployment, and use of AI capabilities.
- Equitable: The department will take deliberate steps to minimize unintended bias in AI capabilities.
- Traceable: The department's AI capabilities will be developed and deployed such that relevant personnel possess an appropriate understanding of the technology, development processes. and operational methods applicable to AI capabilities, including with transparent and auditable methodologies, data sources, and design procedures and documentation.
- Reliable: The department's AI capabilities will have explicit, well-defined uses, and the safety, security, and effectiveness of such capabilities will be subject to testing and assurance within those defined uses across their entire life cycles.
- Governable: The department will design and engineer AI capabilities to fulfill their intended functions while possessing the ability to detect

> and avoid unintended consequences, and the ability to disengage or deactivate deployed systems that demonstrate unintended behavior.

From the start, the participants in Pivot Projects were aware of the ethical questions that their involvement with AI might raise. They agreed early that they would not collect or use any identifiable personal data in their analyses. Beyond that, however, they didn't come to grips for months with the ethics surrounding AI in a structured way. The leaders had charged the Faiths workstream with addressing the issues, but the group didn't produce or adopt guidelines. Members of the Faiths workstream *did* explore the ethical issues in their own discussions and when they joined other workgroups. Sarah Joseph, the journalist who had converted from Christianity to Islam as a teenager, was especially attuned to the topic. "When we work with SparkBeyond, the human element must be fundamental," she said. "We have the faculty of discernment. It's up to us to ask the right questions, to filter out the noise, to be aware of bias, and to recognize what will be a useful insight."

Bias was a particular concern of the Pivot Projects leaders. As scientists and engineers, they were acutely aware of the limitations of scientific inquiry. "You can't validate absolute truth. Some scientists may claim to do it, but science is an ongoing process," said Colin. "Conflict is inherent in the scientific method. Hypotheses compete until we find one that's best, but, even then, it's 'best' provisionally. A better idea may come along."

How, then, would the group deal with bias? They understood that every worldview is biased. They believed that by adopting the SDGs and the Earth Charter and by making Pivot Projects as diverse as possible, they would improve the validity and credibility of their work. Were there other checks on bias that would be viable and useful? In December 2020, the British manufacturer Rolls-Royce Co. released its so-called Aletheia Framework (named after the Greek goddess of trust and disclosure)[5] to the public, stating that businesses that followed the checks and balances within the framework could rest assured that their AI projects are fair, trustworthy, and ethical. This framework could help Pivot Projects set its moral compass.

Colin also sought counsel from experts, in particular members of the International Observatory on the Societal Impacts of AI Digital Technology at the University of Montreal. He, Ian Abbott-Donnelly, and Shay Hershkovitz held a Zoom meeting with a trio of faculty members—Nicholas Martin, Bryn William-Jones, and Christophe Abrassart. The faculty members were not familiar with the SparkBeyond platform, so they had only general advice for the Pivot Projects team: be aware of the biases associated with working exclusively in English and with the fact that much of the Kumu modeling was being done by males, who carry gender biases. They also cautioned about the age-old garbage in, garbage out issue with computing, and Shay assured them that the SparkBeyond system filters untrustworthy sources of information. In the end, the faculty members offered no magic bullet for identifying and dealing with bias. Looking back on the conversation, Colin commented wryly: "Where is absolute truth when you need it? We need some sort of bias police but don't know where to find it."

To get another point of view, I reached out to David Danks, head of the Department of Philosophy at Carnegie Mellon University and the chief ethicist of the university's Block Center for Technology and Society. I gave him the background on the project. He said he was impressed. "This approach sounds better—less biased, more useful, less parochial—than most others," Danks said, yet he cautioned the group to be wary of vulnerabilities that might harm the validity and effectiveness of their results. For one, he pointed out that, while the SDGs are relatively bias free—in the sense that most people would endorse them—the SDG framework provides no guidance about how to resolve conflicts or trade-offs between the goals. "They're not going to be able to eliminate biases or all introduction of their own values," Danks said. "I would argue that the real goal should be for them to be as explicit as possible about the value-laden choices that they were making. Where are the values making a difference, perhaps only implicitly, in the project?"

I circulated Danks's feedback to SparkBeyond leaders and a diverse array of Pivot Projects people and got responses that were all over the map. Stephen Passmore, the president of Resilience Brokers, took the position that all biases are bad because they make it difficult to consider or discover another point of view. Shay Hershkovitz, of SparkBeyond, said all information is biased, so it's futile to try to eliminate it. "Biases can be valuable or not, but they always exist," he said. The important thing, he said, is to be aware of that and to use a combination of human and machine intelligence to remediate the biases that exist in the information we collect, the mental models we use to organize it, and the questions we ask about it.

What Pivot Projects people were left with were the tools of human judgment and machine information processing. Machines provide the capacity to gather and synthesize huge quantities of information and can also serve as a check on biases. Humans provide the judgment to reach useful conclusions based on synthesized information, but they must have the humility to question all of the assumptions upon which those facts and conclusions are based. There are few easy answers in this complex world—at least if you have your eyes wide open.

AI: From the Wings to Center Stage

Over the past decade, AI has moved from the wings to center stage in business and society. It is improving our understanding of what is happening in the world around us and why. Armed with superior knowledge and with better capabilities for predicting the future, government and business leaders, scientists, and a wide range of other users of AI are able to make better decisions and produce better outcomes. As we look ahead to the coming decades, we can take AI techniques that have been deployed successfully by businesses and governments and apply them to climate change.

SparkBeyond's problem-solving platform is one of literally thousands of tools that can be applied to different aspects of the problem. A group in the United States called Climate Change AI (www.climatechange.ai) was formed to apply information technology to problems associated with global warming. It issued a report[6] in 2019 that's essentially a guidebook for using various machine learning techniques to reduce greenhouse gas emissions and helping society adapt to a changing climate.

Among the authors of the report are Yoshua Bengio, winner of the 2019 Turing Award, and Felix Creutzig, a coordinating lead author for the UN's Intergovernmental Panel on Climate Change (IPCC) *Sixth Assessment Report*, the definitive update on the status of climate change, which is due out in 2022.

Climate change is by far the biggest and most complex challenge that humanity has ever faced. It has been said that we already have all the technologies we need to deal with it. That's certainly true in the sphere of information technology and AI. Now it's up to societies and individuals to summon the will to change: to organize, to invest, and to take on this beast. Pivot Projects and the Climate Change AI group are two examples of how a diverse array of people with a wide variety of domain expertise can harness AI to produce practical solutions to climate-related problems. They represent augmented intelligence at its best and potentially at its most fruitful. But these are just two tiny initiatives in a vast sea of humans, most of whom are focused on something other than the survival of the species. Collaborative intelligence and augmented intelligence have tremendous potential, but they can fulfill that potential only if they are put to work on a massive scale.

PROFILE

Shay Hershkovitz

Israeli political scientist and technology executive

When Shay Hershkovitz was sixteen years old and growing up in Tel Aviv, Israel, he used to get up in the wee hours of the morning to watch Boston Celtics basketball games on TV. He loved the team's blue-collar work ethic. Another unusual behavior for an Israeli teenager: he immersed himself in the writings of Karl Marx. His history teacher in school, Haim Grossman, led the kids off the traditional curriculum track into some of the great works of political philosophy, including the early Marxists, the nineteenth-century Liberal thinkers, the anti-Fascists, and even pamphlets from the French Revolution. "It was a life-changing experience for me," Shay said. "That's when I started building my political consciousness. I pulled to the left, in the direction of tolerance and respect for human rights."

Shay's life journey has been guided by those core principles, and he found outlets for his desire to make the world work better in a wide variety of ways, including doing academic research, authoring books, teaching at a university, starting a company, and driving social missions for organizations. He hooked up with Pivot Projects shortly after it launched through his role as Climate Action leader at SparkBeyond, the Israeli artificial intelligence (AI) company. "We're super aligned," Shay said. "Both organizations want to create a tool that's open to the public and can be used by researchers and decision makers from around the world. SparkBeyond can offer the platform, but we can't model everything ourselves. We need the alliance."

Shay typically took a back seat on Pivot Projects all-hands meetings, but he came into his own during regular weekly meetings between the two organizations. He had a frank but friendly approach to expressing himself on Zoom calls. He ran through his checklist of action items like a man who had another meeting to get to, which was typically true, yet he was fully engaged. Whether the news was fabulous or frustrating, nothing gave him pause. If they ran into a barrier, there was always another way to get stuff done.

He had packed a lot of variety into his life. He studied Islam and Arabic in school along with political science. He served in the Israeli military for a decade. He ultimately got a PhD in political science and traveled back and forth between academia and the start-up world for several years before landing in Los Angeles as the head of research for XPRIZE in 2017. This role put him on a new path spearheading innovations that combine data, technology, and multidisciplinary collaborations.

XPRIZE is a nonprofit organization that identifies important challenges to be overcome, raises money from sponsors, and sets up cash-prize contests aimed at solving those problems. Shay's job was to build a team that could quickly gain expertise about issues that the organization might target. He hired a diverse group of analysts—generalists who could take on any topic. The team worked in sprints of lasting two to four months, but he felt the work was too manual. They needed technology that could accelerate the search for expertise and insights. That quest took him to SparkBeyond, where he hit it off with cofounder and CEO Sergey "Sagie" Davidovich. XPRIZE used Spark Beyond's AI machine to automate its research.

At the time, the term *big data* was all the rage in tech circles. Big data was the idea that the vast quantity and variety of digital data that has become available makes it possible for humans to understand what's going on in the world more deeply and to make better decisions. But Shay saw that data by itself wasn't sufficient. A smart melding of data, expertise, and diverse perspectives was needed. He coined the term *big knowledge* to describe that union.

A couple of years later, Shay and Sagie came up with another big idea: why not address the world's most challenging problems by combining SparkBeyond's AI with crowdsourcing, the technique of getting thousands or tens of thousands of people involved in a project that promotes the common good? They set up the nonprofit Climate Action project at SparkBeyond, and Shay joined the company to lead it. His job was to raise money and to develop alliances with groups of experts from around the world who could use the AI to help deal with climate change.

With the onset of the COVID-19 pandemic, SparkBeyond's business came under stress, and the company put the Climate Action plan on temporary hiatus. Then came Pivot Projects. Here was a way that Shay could advance the vision that he and Sagie shared without breaking the bank, and he jumped at it. "Pivot Projects is my dream," Shay said. "We're testing ideas. We're training the machine. We're building the models. We're harnessing human cognition to help the machine do its job better."

9

Points of Light

In 2015, community leaders and residents in the college town of Chapel Hill, North Carolina, came to a disturbing conclusion: the municipality of 60,000 people and the towns surrounding it were way too dependent on the automobile for getting around. Cars were clogging the streets and their exhaust was fouling the air. Meanwhile, walking and biking could be perilous. The situation wasn't sustainable, so the city launched an initiative aimed at reducing car traffic and increasing walking, biking, and use of mass transit.

A plan came together quickly. It included improvements to mass transit and dozens of miles of new bike paths and bike lanes. The goal was to shift noncar transit from the 23 percent mark achieved in 2011 to 35 percent in 2025. A core project team tapped into multiple city and regional agencies; officials from adjacent towns; the University of North Carolina–Chapel Hill; local businesses; and numerous community organizations, including running and bicycling clubs. To reach as many people as possible, they staged pop-up events, conducted web surveys, and held large public meetings. The planning took eighteen months from launch to adoption.[1]

Before the COVID-19 pandemic, the city was on track to reach its goals. About 27 percent of trips were being done using shoe leather, bikes, and buses. "A lot of people here understand the value of pedestrian and bike investments," said Corey Liles, a senior planner for the town. "Our town council gets it, community leaders get it, and developers understand it. The plan has teeth. It doesn't just sit on a shelf. If it's in the plan, it is going to get done."

Problem Solving at the Local Level

Within Pivot Projects, the Engagement workstream spotlighted Chapel Hill's mobility plan as a shining example of how communities can identify problems related to sustainability and quickly do something about them. All too often, urban transformation initiatives take years to get going, set goals that are too grandiose and amorphous, fail to include robust input from people in communities, and aren't monitored closely to make sure they achieve their goals. Chapel Hill avoided those potholes by adopting a highly disciplined approach for planning and execution. It harnessed a small and agile core team, did intensive community outreach, and set concrete goals and actions.

People in the Engagement workstream had a lot of experience with this sort of thing. Most of them were veterans of IBM consulting and/or public sector business units and had been involved in Smart Cities projects. They knew how to deal with government agencies, which typically move slowly and are reluctant to make bold moves. At the same time, they had seen how innovation gets bottled up in big tech companies. All of them had retired from IBM, but they were driven to help society

Gerry Mooney

address climate change. “Our IBMers in this project see the thing over the horizon and it scares them. They’re motivated,” said Gerry Mooney, the leader of the Engagement group who had led IBM’s global government business and Smarter Cities campaign. In addition, there were non-IBMers onboard who brought different perspectives: Bert Jarreau, who had been chief innovation officer at the National Association of Counties; and Julie Alexander, director for technology and innovation at Places for People, a London urban development consultancy.

The Engagement group acknowledged that one mobility project in one small city in one country won’t change the trajectory of global sustainability, but they believed that, if thousands or tens of thousands of communities around the world launched similar programs, it could amount to something. “Every little incremental project is a shining light. A small group of people can step up and get real work done,” said Jim Cortada, one of the former IBMers. “If you have a lot of places doing the same kind of things, it lights up the sky. When you have all the lights on, you have fundamentally changed the world.”

The group published a report outlining their engagement model and rationale, and they created a two-page tip sheet for community leaders (see the end of this chapter). In the late summer of 2020, they began reaching out to communities to test their ideas. They hoped to form partnerships with nonprofit economic development organizations to get their tip sheet into the hands of government and community leaders around the world. Like others within Pivot Projects, they were making it up as they went along, but they were applying lessons from their many decades of experience.

The Pivot Projects leaders had created more than twenty workstreams, each focusing on an aspect of society or an intersection of nature and humanity that they deemed to be strategically important. Still, the workstream topics were quite broad, comprising huge slices of human experience. Choosing key words and creating Kumu models from them laid the groundwork for interacting with the AI platform, but the core task of each workgroup was to survey the landscape of the system they were studying and to pick a spot or two where they felt they could make a difference. “One of the most important things is you don’t try to boil the ocean. Pick some really useful things you can do at the local level, and get it done," said Gerry.

Gerry had spent practically his entire career in the tech industry. Early on, he cofounded a Silicon Valley microchip company and served as its president. Then he spent twenty years in management at Hewlett-Packard and IBM. Since his retirement from IBM in 2014, he had served on the boards of companies and nonprofits and had advised start-ups. He had a lot of experience with innovation in small companies and large ones. At Hewlett-Packard and IBM, it was difficult to turn radical new technologies into businesses, especially if they threatened the organization's existing revenue streams. Still, that kind innovation happened on occasion. "Somebody inside, a world-class expert, would see over the horizon," Gerry said. "They saw something new and important, and they pushed it like crazy, even if it meant they got in trouble for it." He saw that same fighting spirit in Pivot Projects.

Developing a Model for Engagement

The Engagement group had gotten off to a meandering but thought-provoking start, which was quite typical of Pivot Projects discussions. They had started with a different name, Industries and Business, which the project leaders had established as one of the core workstreams because business was a fundamental system in any society. Later, I convinced the members of the group to change their name to Engagements after it became clear what they would focus on. Colin Harrison had recruited Gerry to run the workstream. They had worked together at IBM on the Smarter Cities project.

At first, the group focused on radical innovation by companies and within industries. They latched onto the ideas of Wharton School professor George Day, who had bemoaned the shortfall of bold innovations in U.S. industry. He talked about Little I (incremental innovation) and Big I (radical innovation). He asserted that there wasn't enough Big I being done. The problem was that many organizations had a strong aversion to risk, he thought, and the solution was to distribute research and development (R&D) projects more evenly across a spectrum of risk.[2]

Gerry drew parallels to the COVID-19 pandemic. Would businesses and industries settle for a return to the old normal or would they seek to produce radical innovations that could set the planet on a more sustainable path? One way to make bold innovations less risky was for businesses to work in partnership with governments. That spread the risk, for one thing. But Gerry also believed that unless governments and businesses worked together, there was little chance of making substantial progress. Gerry's proposal: "Let's create a framework to address the change that the public and private sectors need so they can work together to conquer climate change. We can think of this as Little P [pivot] and Big P."

Before the COVID-19 pandemic, one of the biggest failures in California relating to climate change had involved the state's power utility, Pacific Gas & Electric (PG&E). In 2018, the company's poorly maintained electrical grid had caused the so-called Camp Fire, which wiped out the town of Paradise, caused the deaths of eighty-four people, and ultimately drove PG&E to seek protection from bankruptcy. At one point, California governor Gavin Newsom suggested that the state take over PG&E and transform it into a truly public utility. That didn't happen. Gerry wondered whether a more radical approach might have been more successful. It might have been better, he suggested, to bring together the best minds in government, the utility industry, and the Silicon Valley innovation community to develop a new model for producing and distributing energy that would be more sustainable, would create jobs, and might cut down massively on the need for the kind of transmission lines that destroyed Paradise. This was the kind of radical public–private partnership that he thought Pivot Projects should explore.

While public–private partnerships of various kinds had been tried around the world with some success, they were typically bureaucratic and slow-moving beasts. With the COVID-19 pandemic ravaging societies, economies in distress, and climate change happening before our eyes, there was no time for delay. So the group explored a technique that had become popular in the world of design but was also used in urban planning and economic development: the charrette. This is an intense session in which a group gathers in a location (typically separate from the normal workplace) where they brainstorm, develop a collective vision of the future, and map a process for turning their vision into a reality.

A key element of the charrette is that, in spite of the fact that it's done at the beginning of a process, there's deadline pressure. The participants understand that they must get something done in a few hours. That fosters big ideas, the spirit of collaboration, and the willingness to agree on something and commit to it.

Introducing the Virtual Charrette

Gerry had learned about charrettes and participated in them as a trustee for the Urban Land Institute (ULI), a nonprofit organization dedicated to research and education in land development. The ULI organized charrettes in cities that were contemplating big sustainability projects, but it could afford to do only four per year. It was expensive to fly experts in from around the world and put them up in fancy hotels. One lesson that COVID-19 had taught is that groups with widely scattered members could conduct quite effective meetings on Zoom or other video collaboration platforms. So Gerry proposed virtual charrettes, which could be done inexpensively and conveniently. The features of modern video collaboration tools provided everything a charrette required, including screen sharing, sidebar chats, surveying, and breakout sessions. The main difference was everybody involved would provide their own coffee and croissants.

Charrettes typically include a diverse group of participants. One broad category is the subject matter experts—a multidisciplinary bunch. Another is stakeholders. For a sustainable transportation project in a city, the stakeholders might include the mayor, city department heads, regional planning people, heads of advocacy groups, and citizens. To the folks in the Engagement workstream, it was essential to have people who actually do the work or experience the impact of it as participants. They had seen too many situations where leaders decided what was best without consulting the people most directly involved. And it wasn't enough to have listening sessions where community people get a chance to make comments or ask questions about a project. They had to participate in the charrettes as equal partners.

The group referred to this localization imperative as working at the coalface, a term that originated in English mining communities and was based on the idea that the miners who were drilling and dynamiting coal 300 feet below the Earth's surface had better ideas about how to improve mining than did managers sitting in comfy offices in town. To get things done, the leaders of urban sustainability projects needed to listen to and empower local stakeholders. Another dimension of the metaphor was that it was most effective to work at the local rather than the national level and to focus initially on concrete, achievable goals. "By talking about the coalface, we want people to start small, build their confidence, and take on more ambitious things over time," said Gerry.

One telling example of community involvement of this type is the Room for the River program in the Netherlands. After terrible floods across northern Europe in the 1990s, several countries launched initiatives aimed at flood abatement. Rather than just resisting rivers with dykes, dams, levies, and the like, the focus was on accommodating them. Typically, responses to challenges like this are top-down affairs in which central governments decide what to do. The Netherlands took a different approach. While the central government defined the program and provided funding, it was up to local communities to choose the solutions that worked best for them. In some municipalities, citizen groups were invited to propose their own solutions. The program resulted in a mix of approaches that, so far, has worked. Strategies included relocating dykes farther from the riverbank, removing soil and rock to reduce the level of the floodplain, removing obstacles in the rivers, and relocating families and businesses.[3]

The Engagement team embraced Peter Head's concept of the collaboratory—or collaborative laboratory—a mix of people with different skills and expertise working together to solve a problem. These would not be garden-variety charrettes. Pivot Projects would bring its diverse teams of experts, its Kumu models, and the SparkBeyond artificial intelligence (AI) capabilities to the gatherings. The Pivot Projects teams would work arm-in-arm with local leaders and experts, they would customize the Kumu models by adding local priorities and values, and they'd put local data into the SparkBeyond machine. There would likely be a design charrette at the beginning of the process, but then there would be an ongoing series of engagements between humans and the machine—a ton of collaboration and research—culminating in a plan and a process for making significant changes, for making places more sustainable and resilient.

The team believed that, in order for the virtual charrettes to work well, they should be run by experienced facilitators who have deep knowledge of systems thinking, Kumu modeling, and the capabilities of the AI platform.

Lessons from Chapel Hill

Within the Pivot Projects Engagement workstream, Curtis Clark had taken on the assignment of collecting exemplary city initiatives from around the world, but in the end, he chose his own hometown, Chapel Hill. He lives two blocks from the University of North Carolina campus and two blocks from city hall and is a direct beneficiary of the mobility project. He and his wife frequently walk or bicycle around town, and they were thrilled at the improvements. "They redesigned the flow of people and bikes and cars in ways that enhance mobility and keep everybody safe," he said.

One of the keys to the success of the plan was getting input from stakeholders and getting it early. Just one month after launching the planning process, town officials began gathering input by arranging pop-up meetings on sidewalks and along greenways where they could easily interact with walkers and bike riders. A month later, they launched a planning Wiki online. Anybody who wanted to could write comments on the plans as they developed, submit their own ideas, take surveys, and draw on maps.

The town didn't have funding to make all the changes all at once. Instead, it piggybacked on street and sidewalk projects undertaken by the state department of transportation or private developers. Street paving projects offer opportunities to create bike lanes and paint brightly colored crosswalks.

The biggest changes so far were made to East Rosemary Street, which creates the border between downtown and the vibrant Northside neighborhood. Before the alterations were made, East Rosemary was a three-lane road that was often clogged with traffic. You wouldn't go there unless you had to. The town turned it into two lanes and created buffered bike lanes on each side. They widened sidewalks, planted trees, and reduced the speed limit to twenty miles per hour. "The way they designed the corridor changes the safety, the rideability, and the aesthetics. It's a pleasure to be there," said Curtis. On East Rosemary Street, bike traffic more than doubled and the area averaged more than 1,100 walkers per day.

Curtis had worked in state government for many years, serving as the chief information officer for North Carolina before joining IBM, where he spent twenty-two years—all of it in government sales. While at IBM, he traveled the world helping governments on all levels digitize their operations, shift online, and adopt smart-cities technologies. He saw a lot of failed technology projects and a few good ones—which line up well with how things went in Chapel Hill. Here are his three rules for engagement:

- Top political leaders must be committed to the journey. It can't just be a campaign slogan. They must be willing to spend the time to understand deeply the problems they want to solve, to allocate adequate resources, and to stick with it over the long haul. It is best to appoint a transformation agent who has the capacity to coordinate across government departments and interest groups and thus get buy-in from everybody.
- Transformation agents must be good at Small P politics. Most government agencies operate in isolation from one another, jealously guarding their authority, budgets, and data. Change agents have to understand the needs of the department heads, earn their trust, and persuade them to work together. Everybody has to agree on goals and the right metrics for measuring progress.

- Leaders have to think long term. Politicians like to propose grand new projects. That's how they get elected. So major transformation projects, which typically take several years to complete, require a different frame of mind. Leaders must institutionalize their transformation projects so that the good work will continue when administrations change.

Too often, Curtis said, he would get calls from people in cities a couple of years after they launched new transformation initiatives. Things had fallen apart, usually because they had failed to abide by his three rules of engagement. They had beautiful plans but dysfunctional organizations. "You need leadership commitment, governance, and community commitment," Curtis said. "If you don't have those things, nothing else matters."

A New Type of Engagement

From the start, members of the Engagement workstream saw their primary task as creating a tool that Pivot Projects could use to manage its engagements with communities. The tool would be a bridge between the solutions that the workstreams produced and the communities in which they hoped to test their ideas. Gerry, Jim, and their colleagues kept the larger group apprised of their progress and solicited feedback. Jim was the primary liaison. A presentation that he made to an all-hands meeting on July 10, 2020, awakened people to the potential for replicating their collaborative intelligence model on a massive scale: "the thousand points of light," which Jim had cribbed from themes explored by President George H. W. Bush. I mentioned this pivot point in chapter 3.

Jim had an easygoing style that the group responded to. His soft approach to group dynamics had served him well throughout his career, which he had spent almost entirely with IBM. He had been a consultant, consulting manager, and sales leader for government clients for thirty-nine years and had spent more than a dozen of those years researching public sector management and strategy. Somehow, Jim had also

Jim Cortada

found time to write a dozen books, mostly on the history of computing and information technology. While he was involved in Pivot Projects, he was writing a new book about the role of individuals in transforming societal ecosystems. It was as if Jim's whole life had prepared him for this volunteer gig.

At an all-hands meeting on September 4, 2020, Jim presented a summary of the group's report that set off a rapid-fire discussion of styles of engagement. The group members talked most about citizen's assemblies, an idea that traces its roots back to the days of classical Greek democracy but has been updated in the twenty-first century to address the perceived need to get more citizen input at the national and local levels. Members of citizen assemblies are typically selected at random and serve for several years. They generate proposals that must be approved through citizen referendums before becoming law. They have been established in a number of countries, including Canada, the United Kingdom, and Poland. The United Kingdom launched Climate Assembly UK in 2019 and published a proposal based on citizens' input.[4]

At that all-hands meeting on September 4, Deborah Rundlett, one of the most active participants in Pivot Projects, urged Jim and his group to add another dimension to their engagement model: inspiration.

"This change process isn't just about problems to solve but also about identifying the aspirational goals that communities want to reach," Debbie said. "We begin with the idea, the vision. It engages a different part of the brain. We don't ignore problems and problem solving, but we engage first around a shared vision." Jim agreed. He promised to weave that into the narrative. He pointed out that people in communities often want to talk first about their problems. They feel compelled to get their complaints off their chests, but they also want to address their dreams of a better future for themselves and their children. "We need to signal the positive approach," Jim said. "We can have a better climate, we can have a better economy, and we can have better health."

Thinking positive never hurts. And neither does creating a crisp and clear guidebook for people to follow. But there are a lot of client engagement models. Every consulting firm has one, and sustainability firms like Peter Head's Resilience Brokers have used theirs extensively in their work with communities around the world. Could the Engagement workstream's approach make a difference? Could community leaders take a broader view in the midst of the COVID-19 pandemic? Also, it was important to recognize the difference between a Pivot Projects engagement with a city and the way IBM approached such things. IBM's consultants could work with city leaders and residents to map a project, then the company would provide the consulting, computing, and software that was needed to execute on the plan. It was a commercial transaction with a $100 billion company standing behind it. Pivot Projects was an ad hoc group of volunteers with no service delivery mechanism as yet. "Building and making it work are the hard parts," said Colin Harrison. "The world is full of books and articles about how to fix the planet, but we need more people in the trenches actually fighting the battle. How can we mobilize a Green Army?" That question needed answering.

Still, the Points of Light model was appealing. It could be a catalyst for change. Communities had been swept up in a global crisis of multiple dimensions. There was a sense of urgency. Many of them were desperate to find a path forward to a less stressed and more just world. In the weeks ahead, Pivot Projects would discover whether what it had to offer was appealing to communities under stress.

Tip Sheet

Pivot Projects: How People and Communities Can Organize for Sustainability

Purpose:
The purpose of this methodology is to help groups of individuals in any size community address local economic, environmental, and social problems to be solved following the end of the COVID-19 pandemic. It rests on several assumptions:

- That the post-pandemic economic recovery will require the engagement of many groups of citizens, officials, and experts working together.
- That lessons learned during the pandemic can be applied to economic recovery and ongoing sustainability.
- That large swaths of a society can address local problems.
- That every community possesses local pockets of leadership needed to execute the approach to resolving local issues.
- That there are proven methods to accomplish worldwide local initiatives to deal with such issues.

Approach:

- Define projects that can be completed, say, in one year, and can then be followed by another one, and then a third. Small teams meet to identify and shape initiatives.
- Consume a "committable" number of resources that can be budgeted by a local government or that can be raised through contributions or other means.
- Pick projects that are replicable across similar communities, such as other towns or cities, so that others can learn from the experiences of others and thus save time and money.
- Collaboration is key among all the stakeholders in a project and community with the objective of keeping a project from becoming

too complex. That means having a single leader, clearly identifiable accountability, and achievable objectives and measures of progress.

- Identify the desired local impact but also the national and global impact. When looked at together, many projects have broader impact both nationally and globally.
- Use need for change to go green to improve the world's environment.
- The methodology is called a charrette, which is a short, collaborative meeting during which members of a project's various teams collaborate quickly to sketch designs and explore ideas and solutions.

Why:

- Because we have to do much in a short period of time: "We are on the brink of missing the opportunity to limit global warming to 1.5°C."
- Because solving problems requires a broad set of participants and organizations.
- Because incremental solutions, along with broad national initiatives, cumulatively lead to quick, measurable results that make sense to a community.

How:

- First, assemble a team.
- Second, define thoughtfully what problems to solve.
- Third, craft specific local solutions to those problems.
- Fourth, define ways of measuring progress.
- Fifth, implement a plan for getting the work done.

PROFILE

Gamelihle Sibanda

Zimbabwean technical adviser to the United Nations

Gamelihle Sibanda, or Gama, as he likes to be called, has faced plenty of challenging situations during his twenty-four years as a UN technical adviser working in Africa. His assignments took him to more than one dozen countries, a number of which were suffering severe social or climate stresses. None, however, was more difficult than Somalia. Gama was part of a UN team sent there in 2006 to help pick up the pieces after sixteen years of chaotic misrule by warlords following the collapse of the government in 1991. The team helped reestablish government institutions and rebuilt roads and buildings using labor-intensive approaches that put money in peoples' pockets.

As difficult as the reconstruction of Somalia has been, Gama acknowledges that the task of saving humanity from climate change is much more challenging. Helping to set up a new government in an ungoverned society is easier than changing unsustainable practices by existing governments, industries, and people. He believes Pivot Projects and other groups hoping to set the world on a better path will succeed only if they try something new. “Pivot Projects will not make an impact by confronting people who are operating unsustainably, but by creating a new model that renders unsustainable practices obsolete,” Gama said.

Gama joined Pivot Projects because Peter Head invited him. They met when they were participants in the Global Solutions Sustainable Infrastructure Lab, a program run by the Global Solutions Initiative, a think tank focusing on the climate crisis. When it came time for Pivot Projects to pick somebody to manage the approaches to cities and other places, the leaders tapped Gama for the job. He saw himself as an advance scout who would explore relationships around the world and, when real possibilities for collaboration developed, hand them over to a larger team.

Gama spent his preteen years in a rural part of Zimbabwe, where he herded cattle on grasslands bordering areas where wild animals roamed. He appreciated the importance of living in harmony with nature. He later got a degree in civil engineering and went to work for the United Nations. He now regrets that he spent years living in cities and “unlearning” about nature. (Gama now lives in South Africa.) In 2009, he watched a television program about biomimicry. “It lit me on fire,” Gama said, and it brought him back to nature. He sought out Janine Benyous, the reigning world expert, and studied at her Biomimicry Institute in Missoula, Montana, and also obtained a masters degree in biomimicry from Arizona State University.

Biomimicry is the science of using lessons from nature to solve human problems. Gama believes that only by applying biomimicry principles will the environmental movement be able to win major battles in the war on climate change. He has used biomimicry to solve problems in his sustainability work. For instance, in Durban, the third-largest city in South Africa, he was part of a team from BiomimicrySA, an affiliate of the Biomimicry Institute, which was called in to resolve an impasse. Developers wanted to build a massive commercial and residential complex. Municipal officials were concerned that this project would interfere with wetlands that were essential for flood attenuation. The BiomimicrySA team studied the problem and proposed a solution: Why not measure all of the flows of water in the existing natural system (including run-off, groundwater infiltration, and evaporation) and then design the built environment so it mimics those flows? "We proposed that they use nature's ecological performance standard to guide sustainable development," Gama said.

Perhaps the most important use of biomimicry is in designing regenerative solutions to human problems. Organisms heal themselves through processes that make them resilient to damaging changes in the environment. Human systems can also be made more resilient by conserving, recycling, and reusing resources. The principles of regenerative design are used in agriculture, community planning, economics, and ecosystem restoration. Gama pointed to an example of regenerative design in Africa. Lakes and reservoirs have been clogged by an invasive species, the water hyacinth, which disrupts the natural ecosystem and limits fish farming and recreation. Then scientists discovered that the plants' roots naturally absorb pollutants, including lead and mercury as well as some carcinogenic organic compounds.[1] Several African communities are now using the plants for some aspects of wastewater treatment rather than using traditional technology-based solutions that use more resources and are less effective.

Gama came to appreciate fully the entanglement of human problems with nature's when he was working in Somalia. He spoke to a former Somali pirate, who had made money by hijacking ships off the coast of the country and holding them for ransom. The former pirate explained to Gama that he and his colleagues had been coastal fishermen before the country's collapse. Afterward, ships from other nations took advantage of the lack of governance and began dumping waste and toxins in Somali waters, which killed the fish. The local fishermen adapted to their loss of livelihood by pirating ships that plied their waters. Gama said, "You can see how a social problem became an environmental problem which became a livelihood problem which was transformed into a criminal activity."

It was just a short while later that he discovered his calling in biomimicry and set off on his mission to save the planet. It consumed him for more than a decade. But during the COVID-19 pandemic and through his involvement in Pivot Projects, he had an epiphany. "It was my aha! moment," Gama said. "I realized the planet doesn't need to be saved. We need to save *ourselves* from extinction. If we went extinct, the planet would heal quickly."

10

Places

On September 28, 2020, a subset of the Pivot Projects volunteers met on Zoom to consider the next phase of the project's journey. The group had spent months organizing itself, exploring ideas, developing themes and proposals, writing reports, creating Kumu maps, feeding the SparkBeyond artificial intelligence (AI) platform, and developing a model for improving sustainability in communities. Now it was time to start engaging with people in places.

The folks on the call had a wide range of experience with engaging clients. The former IBMers—Colin Harrison, Gerry Mooney, and Jim Cortada—had fronted for a giant corporation selling technology and services to government agencies and companies. The people from Resilience Brokers, Peter Head and Stephen Passmore, had helped cities and regions around the world plan sustainability projects. Damian Costello, a business consultant, had advised large corporations on fostering innovation. Gamelihle (Gama) Sibanda, a UN technical adviser, had helped communities in Africa build infrastructure and create jobs. Ian Richardson had worked within village governments in the northeast of England. But none of them had ever dealt with a situation quite like this before.

Members of the group had already put forth ideas about how to work with places: the Engagement workstream's virtual charrettes and Peter Head's collaboratories, which I described earlier. The two ideas were complementary. Both envisioned close, ongoing, collaborative engagements combining a wide variety of global experts and local stakeholders. The group already had in place the collaboration technology platform they planned on using with communities—the combination of Zoom, Slack, Trello, and Google Docs that was the backbone for their own internal communications. It could be opened up to others with little additional effort. They would invite people from places of interest to collaborate with them to localize the content of the topical Kumu maps and SparkBeyond models. "The question," Colin said, "is how do we make all of these pieces work together?"

The discussion that followed really took off when Gerry, who lived on the northern California coast, raised the specter of the wildfires that burned across the American West in 2020. At that point, the fires had burned more than 3.8 million acres in California—by far the worst year on record. On September 28 alone, 18,700 firefighters were battling twenty-seven major fires.[1] Scientists and political leaders in California had connected the fires directly to climate change. Gerry pointed out that, just a few days earlier, California governor Gavin Newsom had issued an executive order requiring sales of all new passenger vehicles to be zero emission by 2035. The governor's order exposed a conundrum facing California policymakers. The switch to zero-emissions vehicles would result in an estimated 35 percent reduction in greenhouse gas emissions and an 80 percent improvement in oxides of nitrogen emissions.[2] At the same time, however, it might require construction of additional electricity transmissions lines, which had caused a number of devastating wildfires, including the one that devastated the town of Paradise in 2018.

Gerry had a solution in mind: "We need to convince town after town to move from getting electricity from the power grid and move aggressively to local solar. If everybody moves, we can get to the tipping point where we don't need so many distribution lines." Out of that came a discussion of how to approach towns, the state government, and the alternative power industry to combine forces to help towns switch to alternative energy and microgrids.

Add California towns to the lengthening list of places the Pivot Projects had interest in helping to make the transition to improved sustainability and resilience. Members of the group had already reached out to people in a number of places around the world, including Wales; London; Exeter and towns in the northeast of England; Medellín, Colombia;

Montreal and Vancouver in Canada; Delhi, India; and Tribhuvan University in Biratnagar, Nepal. It seemed like every week, another location or two was added to the list.

Every place would be unique, with its own set of problems, priorities, and political dynamics. The Pivot Projects crew was acutely aware that their task would not be as simple as engaging with places, analyzing their problems, and applying best practices to solve them. That may work for corporations in need of a reboot but not for communities seeking fundamental transformation. In some cases, the places they engaged with would be facing crises and deep uncertainty. The crew took guidance from Dave Snowden's Cynefin framework,[3] where he divided the universe of situational possibilities into four categories, each with its own solution pathway.

In situations that Dave defined as complex or chaotic, the Pivot Projects volunteers saw that they would have to leave their usual confidence

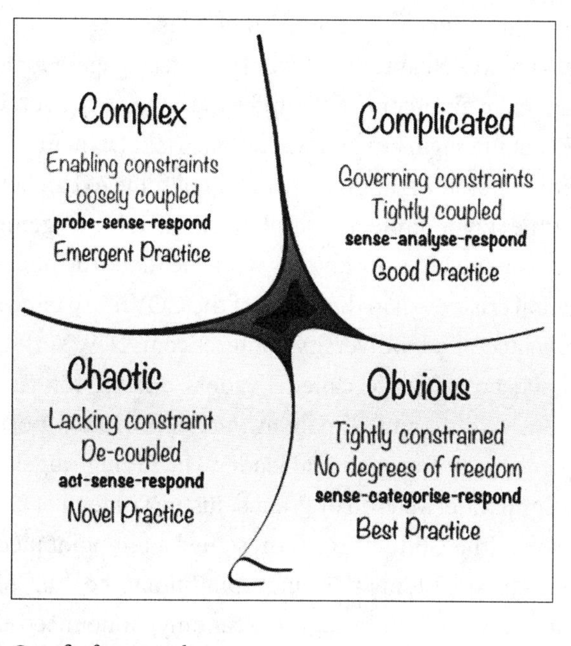

Cynefin framework

Dave Snowden

in the solvability of problems at the doorstep and enter with open eyes and open minds. "The cookie-cutter approach isn't going to work here," said Damian. "We can't expect to cut and paste solutions that work in one place into the next place." Instead, they would have to be flexible, experimental, and bold.

In complex situations, it's difficult or impossible to connect causes and effects directly. So, Dave preached, the best approach is for problem solvers to try a number of experiments in parallel, searching for the solutions that seem to work, then double down on those. In chaotic situations, it's difficult to understand what's going on, much less identify causes. Also, speed is of the essence. Something terrible is happening, and it has to be addressed as quickly as possible. Dave thought that it's best to act immediately with an intervention that seems likely to have a positive effect, monitor the response, and try other experiments if the first one doesn't work. In the COVID-19 pandemic, for instance, hospitals initially put seriously ill patients on ventilators quickly, but they discovered that the approach didn't work well, so they backed off and used drug therapies to reduce symptoms in many cases.

The message was clear for the Pivot Projects engagement leaders. "If we really are in an emergent space and experiencing a pivot, we have to get a handle on managing experimentation," said Damian. "We need to use a systems-of-systems approach, and, hopefully, AI can help."

At the time Pivot Projects was launching its engagement effort, momentum was picking up among world leaders for addressing the environmental crisis even in the midst of the COVID-19 pandemic. That week, in advance of a biodiversity conference in New York City, world leaders signed an initiative called Leaders' Pledge for Nature, committing to put wildlife and climate at the heart of their postpandemic recovery plans. Sixty-four national leaders, including Angela Merkel of Germany, Emmanuel Macron of France, Justin Trudeau of Canada, and Boris Johnson of the United Kingdom, signed a ten-point pledge to help restore the natural systems that underpin human health.[4] Meanwhile, Japan, with the world's third-largest economy, announced the goal of becoming carbon neutral by 2050. Conspicuously absent was Donald Trump of the United States, who had foresworn global collaboration aimed at repairing the natural environment. But he was lagging challenger Joe Biden in the polls. It seemed as if the world could be on the verge of a massive pivot in a positive direction.

The Pivot Projects Pivots

From the start, the Pivot Projects leaders had been talking about the importance of what they called testbeds—engagements with cities and regions. Like raising money, however, this part of the master plan was achingly slow to develop. Some cities already had climate change strategies underway. They didn't need help. Others hoped to embark someday, but in the midst of the COVID-19 pandemic and with municipal budgets under extreme pressure, they didn't have the money or the capacity to launch a major long-term initiative—not even to address something as critical as the survival of the species. Delays in developing SparkBeyond's AI platform for this purpose also held the group back. Yet even as COVID-19 infection rates began to climb steadily again, the group started to gain traction. It felt a little bit like the early moments of a classic roller-coaster ride. You're pulling out of the boarding dock, gradually picking up speed, climbing, climbing. You're looking down as the garishly lit carnival scene gently recedes. The anticipation builds for sudden acceleration and, perhaps a wild, exciting ride. It wasn't clear yet that that last part would actually happen for Pivot Projects, but the signs were strengthening. Peter was thrilled. "This is a pivot. We're shifting the organization. We're moving it out of the research world and into the real world," Head said.

By mid-October 2020, members of the group were reaching out octopus-like to cities around the world. Gama and Peter had a Zoom meeting with a dozen officials from Delhi, India. The city was developing its master plan for 2040. Stephen Passmore was in discussions with people in Glasgow, Scotland, which was to be the site of the twenty-sixth United Nations Climate Change Conference of the Parties (COP26) meeting in November 2021. Ian Richardson was talking to Jamie Driscoll, the mayor of his hometown, North of Tyne, in the north of England.

Peter and Colin had spoken to Tim Lenton, director of the Global System Institute at Exeter University, who had agreed to set up a living laboratory program there for graduate students working on independent study projects in Exeter. They were also talking to people at Anglia Ruskin University in Essex about doing something similar there. Gama and Paul Quaiser, a new addition to Pivot Projects, meet with people at Waikato University in New Zealand about sustainable development in the city of Hamilton. Another newbie to Pivot Projects, T. Luke Young, had put Gama in touch with city leaders in Medellín, Colombia. Aneta Popiel, head of the Construction workstream, was reaching out to London. Colin and Ian Abott-Donnelly were talking to medical researchers at Yale University about potentially using the SparkBeyond (AI) platform to help guide vaccination distribution strategies in the United States. Damian was talking to city leaders in Limerick, Ireland. Colin had met on Zoom with faculty members from Tribhuvan University in Biratnagar, Nepal. Exploratory contacts had been made at a half-dozen other places, as well.

Suddenly, there were lots of irons in the fire. There was no assurance that many—or any—of them would turn into real engagements, but Pivot Projects had to get its act together. The ad hoc style of the first months had worked well, but now the group had to figure out how to approach potential partners with a crisp message and a sound process. During an all-hands meeting in mid-October 2020, Colin sketched out a plan:

- For each testbed there will be a Pivot Projects team leader drawn from the workstreams and with expertise matching some of the known needs of the testbed.
- The leader will invite other Pivot Projects members from the workstreams as needed to broaden the pool of expertise and cover the testbed's needs.
- The testbed partners will likewise identify a leader and one or more members to join Pivot Projects.
- These collaborative teams will develop plans and funding proposals and eventually evolve into the joint research teams.

At the same time, Pivot Projects and SparkBeyond had formalized their relationship. It was a handshake arrangement, with no lawyers involved. When the time came to cut a financial deal with a testbed city to pay for computing time and advisory services, SparkBeyond would be the official service provider and Peter Head's Resilience Brokers would be a subcontractor providing consulting services. Will Maunder, SparkBeyond's sales leader in the United Kingdom at the time, was completing the paperwork so they could participate in the country's AI/climate fund. This would be a hybrid model, with some services paid and others provided on a pro bono basis. As far as the Pivot Projects leaders knew, nothing like it had ever been tried before.

Gama was in the thick of managing the outreach. He had a day job working for the United Nations, but in his spare time he developed a process for tracking the progress of engagements. Pivot Projects volunteers were free to reach out on their own and try to get conversations started, so Gama's management system—essentially a Google Docs spreadsheet—was designed to capture what was happening. The spreadsheet captured all the basics: places, names, contact information, descriptions, to-do items, and status updates. He pulled an all-nighter to prep the spreadsheet and make an explanatory video, and he finished the video just one hour before the all-hands meeting on October 16, 2020. Said Gama, "With this, we'll be able to prevent information overload."

Engaging with Exeter

Tim Lenton had been enthralled with complexity practically since he was in short pants. As a teenager growing up in England in the 1980s, he fell in love with physics and devoured science books like they were popcorn. And when he arrived as an undergraduate at Cambridge University, he feasted on Mitchell Waldrop's new book *Complexity: The Emerging Science at the Edge of Order and Chaos*, about the Santa Fe Institute and the new science of complexity. He was also drawn to Jim Lovelock's writings on the Gaia hypothesis, where Lovelock argued that Earth functions as a self-regulating system, similar to a living organism. Reading about the

Gaia hypothesis sparked Lenton's passion for studying the planet as a whole system, which culminated in his PhD thesis about the factors that regulate the nutrients in the ocean and the oxygen content of the atmosphere. His intellectual journey ultimately led him to Exeter University, where he established the Global Systems Institute, which uses complexity theory and systems modeling to solve sustainability challenges and gives students hands-on experience in problem solving. Suddenly, in mid-2020, the world was in crisis, and people in government and business were beginning to understood that complexity theory wasn't some arcane academic pursuit. It could help them their find their way out of chaos. "I had grown up with this stuff," Lenton said. "But these times we're in brought the rest of the world around to this way of thinking."

The folks from Pivot Projects caught up with Lenton at Exeter. They had spotted the city of Exeter as a potential target for collaboration, but when they made advances to municipal leaders, they were told that COVID-19 was all consuming and there was no time for anything else. Somebody in Pivot Projects learned what Lenton was doing and passed the word to Gerry Mooney and Jim Cortada from the Engagement workstream. They decided to pursue an alliance. It took a number advances over a couple of months, but at last they landed a Zoom meeting. To prepare, they read a paper Lenton had published earlier in the year.[5] Lenton and coauthor Simon Sharpe from the UK government's Cabinet Office, had argued that, in complex systems including human society, tipping points can occur when small, strategic changes result in a cascade of positive changes at ever-larger scales. They believed these "upward-scaling tipping points" could accelerate progress in addressing climate change. They cited as an example the cooperative efforts by small coalitions of governments around the world mandating shifts to electric vehicles. Lenton's paper was one of the more hopeful scientific papers concerning climate change in recent memory. It struck a chord with Gerry and Jim For his part, Lenton saw Pivot Projects as an example of a small tipping point that had the potential to cascade into something bigger. It looked like a good match.

One Zoom meeting led to another and before long, Lenton and Pivot Projects had a handshake agreement to work together. The idea was to create a living lab in Exeter tying together Pivot Projects, the university, the city, and other interested organizations. Pivot Projects would welcome Lenton and some of the students onto the virtual collaboration platform. For their internships, the students could choose among a handful of projects that had been targeted by a public–private partnership called Exeter City Futures. They would tap into Pivot Projects' expertise and the AI platform to develop solutions to local problems. Among the projects that Exeter City Futures had prioritized were a rethinking of the city's rail hub and expanding mass transit into the outlying areas to cut down on auto traffic and fossil fuel consumption. The Pivot Projects people, always thinking big, hoped that similar programs could be established at a number of universities. "We need to train the young people in systems thinking and problem solving," Gerry said. "We need to build a cadre of people who will take this approach around the world."

Jumping on Glasgow's Bandwagon

Glasgow, Scotland, had established itself as one of the most forward-looking cities in the world even before it was chosen to host COP26, the United Nation's climate conference. In 2013, the central government had provided £24 million in funding as part of its Innovate UK program, which Glasgow spent on smart-city projects. It ran pilot programs aimed at reducing energy costs through automated streetlights that adapt to people's activity, provided smartphone apps for around-town travel planning, and combined data from disparate government agencies so they could collaborate better.[6] The central government awarded the city another £150 million in 2017 so it could roll out the initial projects citywide and explore new ones. One focus was on the Digital Glasgow Strategy, a five-year plan launched in 2018 aimed at harnessing the latest technologies to improve the city's economy and the lives of its residents.[7] A year later, the city council announced the goal of becoming the United Kingdom's first net-zero-carbon city. The folks at Pivot Projects saw Glasgow as a prime target. It would be great to showcase their ideas in the place where the world's environmental leaders would assemble as the COVID-19 pandemic, hopefully, began to fade and climate change loomed ahead.

Pivot Projects found two doorways into Glasgow. Nigel Topping, the UN High Level Climate Action Champion, introduced the group to city leaders. Another lead took them to Climate Ready Clyde, an initiative fostering sustainability in the metro region surrounding the city, which is on the banks of the River Clyde.

The direct-to-government route opened up quickly. The first meeting was a get-acquainted session. It was attended by Glasgow city leaders; folks from Pivot Projects; Shay Hershkovitz of SparkBeyond; and Alex Joss, Topping's right hand for technology. Pivot Projects cofounder Rick Robinson knew Glasgow well. When he worked for IBM in the Smarter Planet days, he had run a project in Glasgow aimed at improving heating in the winter for poor people. In fact, Glasgow had already done so much in the open-data and smart-cities areas that the trick was finding something new and exciting that the city and Pivot Projects could do together. People from both sides talked about what they had done and what they might be able to do. Then Gavin Slater, the city's head of sustainability, said, "A lot of this sounds familiar. We don't want to do something we've tried before."

They discussed some of the projects Glasgow had coming up—one investigating universal basic income and another looking at the impact of robotics and AI on jobs. The general feeling was positive. It was worthwhile exploring further. Joss ended the meeting on an upbeat note, saying, "It feels like some magic could start to happen here, though it's not fully clear where."

Looking back at the meeting a few weeks later, Stephen Passmore, CEO of Resilience Brokers and one of the contacts for Pivot Projects outreach, was a little frustrated. He had been doing this kind of work for a decade. "I felt we didn't have a clear message about what we could do for them," Stephen said. "When we go into these meetings, we have to be crystal clear about the offer."

The next session a month later featured Shay Hershkovitz and Ian Abbot-Donnelly demonstrating the capabilities of SparkBeyond's

AI platform by showing an exploration of Ian's water-pricing model, which I explained in chapter 8. Colin was hopeful that something useful would result. "The challenge for us is to show that we can do something different and better than they already do. It's good to be challenged," Colin said.

It looked like this was the way things would go as Pivot Projects attempted to engage with the world. The group's advance scouts would make contact. Introductions would be made. Possibilities would be explored. Demonstrations would be given. There was no assurance that even the most promising matchups would advance to the next stage.

Searching for Truth with Yale

While Pivot Projects was focused mostly on helping governments and communities become more sustainable and resilient, the group also pitched in to help with COVID-19 pandemic management when opportunities emerged. One such engagement—with experts at Yale University's School of Public Health—was aimed at helping policymakers deal with one of the biggest challenges facing healthcare leaders when vaccines arrived for preventing the spread of the disease: the so-called anti-vax movement.

Around the world, many people were reluctant to accept vaccinations to protect themselves and their children from infectious diseases. Since the first such inoculation was given to Russia's Catherine the Great in 1768, vaccines have been an essential part of medicine, critical for eradicating polio and smallpox and slowing the spread of influenza, measles, and pneumonia. In the early twenty-first century, however, vaccine hesitancy gained momentum, driven by peoples' concerns about infringement of civil liberties, religious objections, and health side effects—which were unfounded.

During the COVID-19 pandemic, a significant percentage of the U.S. population doubted the seriousness of the disease because politicians made false statements and media organizations aligned with them published articles containing misinformation. The fear in late 2020 within the medical community was that, because of a breakdown in trust, not enough people would accept the COVID-19 vaccines to achieve herd immunity, and therefore the disease would continue to spread rapidly. Faculty members at Yale were looking for techniques that would quickly identify articles and social media postings that contained misinformation about vaccines so governments and health authorities could counter them and so individuals would know when they were being misled. The spread of misinformation was so prevalent at the time that it had been given a catchy name: the infodemic.

The contact at Yale was Saad Omer, the director of the Yale Institute for Global Health. Omer thought of the idea of producing a truth-serum technology for the internet. A Zoom call on November 2, 2020, bringing together people from Yale, Pivot Projects, and SparkBeyond had started off as a wide-ranging discussion about the possible uses of AI technology to address vaccine hesitancy, but quickly Omer focused on the quality of information. "What we need is quick, automated, high-throughput fact-checking for everything that's on the Web. That's the Holy Grail," Omer said. "People should be able to click on an article or social media posting and know at once the likelihood that it's fake." It was truly a big idea. Ian Abbot-Donnelly got it immediately. "We could start with vaccines but it could be used for anything on the web," he said. "Remember, you heard it here first," Omer joked.

Ian and the team at SparkBeyond set out straightaway to explore the possibilities of developing what Omer was looking for. Shay asked a couple of SparkBeyond's top technologists to lead the exploration. In a couple of weeks, they felt they had the answer: yes, it could be done.

On a November 20 Zoom call, Shay and Ian came back to the Yale team with a proposal: with input from Ian and the Yale faculty members, SparkBeyond would develop a web browser extension that would help readers identify misinformation and guide them to resources containing accurate information. Using a combination of automated web crawling and human curation, they would create a list of false information about vaccines in general, and the COVID-19 vaccines in particular, then use machine learning to train the computing system how to spot such misinformation. Based in part on analysis of the sources of the misinformation, they would enable the computer to assess the likelihood that a statement or the essence of an article or posting is false. It wasn't a real-time fact checker per se. It wouldn't instantly rate a statement as false, but it would alert readers that an article or statement might contain false information and then lead them to more accurate information. Shay said that he thought they could produce proof-of-concept technology in a few weeks and, if it looked promising, produce a working browser extension with a two-month sprint. Speed was of the essence because it appeared that the first COVID-19 vaccines would be rolling out on a mass scale in a matter of a few months. "My concern is that we'll have vaccines but not mass vaccinations," said Omer.

As with all technology innovations, there was no assurance that the capabilities this group of people had proposed during a couple of Zoom calls would actually be do-able or that the technology would be embraced and adopted widely. In fact, after several more weeks of exploration, the SparkBeyond team came back with the recommendation that they focus initially on optimizing Research Studio for researchers, policymakers, and public health communicators so they could better monitor the flow of misinformation and develop strategies for countering it. The browser extension might be too much for SparkBeyond to bite off. But, still, the frisson they had shared was palpable.

It reminded me of the day that Google founders Sergey Brin and Larry Page had visited *BusinessWeek*'s offices in New York City in 1999, when I was the magazine's software editor. This was just a few months after they had spun the company out of a Stanford University lab. They had demonstrated their new search engine to a room full of writers and editors. I was impressed. After the meeting, though, I told a colleague that they seemed like really smart guys and their search engine seemed nifty, but "I wonder if the world needs another search engine." I failed to understand the breakthrough importance of their PageRank algorithm, which ranked search results based on the popularity of the web pages, in other words, how many other articles contained links to them. I didn't foresee that the company would be able to build the most popular website and the most powerful advertising platform in the world and then be able to sell data it collected about users of its services.

Now, after sitting in on the meetings with Yale, I saw that the tech world, and thus the whole world, might be on the cusp of a significant new capability, one that might help not only with vaccine acceptance and ending the COVID-19 pandemic but also with an even bigger problem: the spread of lies that was poisoning politics, social discourse, and comity. This potential breakthrough was made possible by collaborative intelligence, the ability of humans with different types of expertise to come together and augment their thinking with AI. Google had put the world of information at people's fingertips, but with no regard for its veracity. This technology might do one better: help us judge what is credible and what is not.

Gut-Check Time

Just when the Pivot Projects began to gain traction on some of its engagements, discouraging news came from the scientific front. Norwegian researchers Jorgen Randers and Ulrich Goluke published an article in the journal *Nature*'s *Scientific Reports* where they studied current climate data using Earth System Chemistry Integrated Modeling (ESCiMo), a data modeling technique, which had been developed by a consortium of research organizations. They concluded that Earth has already passed the point of no return for climate change. They focused on melting permafrost in the far northern regions of Siberia. Because of climate change, the permafrost was melting on a massive scale, which released large quantities of methane, a potent greenhouse gas. The methane release further increased Earth's atmospheric temperature, which melted more permafrost and released more gases.[8] You see where this is going. It was like a fireball rolling downhill.

At the Pivot Projects all-hands meeting on November 19, 2020, Colin explained the implications of the study's conclusions. The most important question was, Did this new hypothesis mean Pivot Projects should change its focus or direction? If Earth had passed the tipping point, was it still worthwhile to try to mitigate climate change, or should Pivot Projects focus on helping society to adapt to it? As was customary, the participants in the Zoom call broke up into smaller groups for twenty minutes of discussion, then they rejoined the larger group and reported what they had said.

The mood was somber when they reassembled. The consensus was that they needed to keep pressing on both fronts: helping communities become more sustainable yet also helping people and places adapt better to the advance of climate change. So many complex issues intersected here, the whole of earth and humanity, it seemed. It was difficult to sum up what they thought and felt. John Thomas, a retired IBM researcher, gave it a try. "I think part of the problem is the hyper-competitive nature of Western society, where people think of themselves as being this body that's inside this skin, as an individual. You want to optimize your own life regardless of what it does to the rest of the system," John said. He stopped for a minute and gave Zoom a wry smile. "On the positive side, there are things like global consciousness and the fact that people are communicating across the planet that seem to be working toward a more cooperative mental model, which may be necessary for us continuing to have a viable planet."

As sometimes happened with Pivot Projects, the conversation was circular and inconclusive. Meanwhile, the time set aside for the meeting had nearly run out. Then Paul McLean, the Canadian environmental impact consultant, put a cap on it. "If there's a message in all of this, it's that you have to act," Paul said. "You have to just keep doing what you're doing and hope for the best." Andre Head put on some peaceful piano music in an effort to calm everybody down and, one by one, people said their goodbyes and signed off.

PROFILE
Paola Bay

Italian artist and designer

When Paola Bay was six years old, her family traveled to India on vacation. They were a well-to-do family from Milan, Italy. She remembers vividly the chaos of the streets in Bombay, as Mumbai was called then. People, dogs, and cattle all mingled by the thousands. At a giant outdoor crafts fair, her eyes were glued to the colorful textiles, clothing, and shoes and to the faces of the people who had made them. In later years, there were more trips to India and also to Nepal, the Maldives, and Sri Lanka. "The adults were sometimes shocked by the chaos, but for me it was fun," Paola said. "I loved exploring our planet, seeing the different cultures. I love to go to faraway places where Westerners don't usually go."

It's no surprise that Paola grew up to be an adventurer. Throughout her adult life, she has explored the world and herself through travel. The same goes for her professional life. It has been a multi-destination journey. She started as a producer of independent films in London in the 1990s; then, because she had long been a collector of antique *couture* dresses, she became a consultant to the fashion industry; then she became a designer of high-fashion shoes, with her brand, Zoraide. Madonna wore a pair of her shoes in the "Girl Gone Wild" video.

With each chapter in her professional life, Paola explored a domain, accomplished what she wanted to accomplish, and then moved on. Now, she is focused on learning about indigenous cultures and sharing her knowledge with others in the so-called developed world.

And that's how she found her way to the Pivot Projects. She, Peter Head, and Jeff Newman were members of a group in the United Kingdom, the Extinction Rebellion Eldership Circle, affiliated with the Extinction Rebellion climate change movement. Peter had invited her to join Pivot Projects at the beginning and urged her to introduce the group to the wisdom of indigenous people. She was the organizer of the group's webinar featuring three indigenous grandmothers (described in chapter 3).

Paola became extremely aware of the damage humanity is doing to the planet when she began learning about and meeting indigenous people five years ago. As a young person, she had studied Latin, ancient Greece, and philosophy, but now she began to value nature more that the edifices of Western culture. She was puzzled in 2019 when millions of dollars were committed to restoring the fire-damaged Notre Dame Cathedral at the same time that the Amazon rainforest was being burned by ranchers clearing land to raise cattle.

Paola had begun her pursuit of knowledge about indigenous people after a friend introduced her to the Mamos, the spiritual leaders of the Arhuaco people from the Sierra Nevada de Santa Marta rainforest in northern Colombia. The Arhuacos preserved their culture in spite of colonialism and modern times mainly because of the inaccessibility of their mountainous homeland. In Arhuaco culture, the Mamos are responsible for maintaining the natural order of the world through songs, meditations, and rituals. The Mamos prepare for their lives as shamans by spending their teenage years living underground in caves—where they learn to listen to the silence of the earth. Paola has visited them in their villages a number of times and even participated in their ceremonies.

She believes that indigenous spiritual leaders have a lot to teach modern people about humanity's place in the world. In fact, one of the Mamos, Dwawiku Izquierdo, issued a statement about the coronavirus:

> We were so powerful that in a blink of an eye we overheated the planet, thawed the poles, causing many brothers of flora and fauna to disappear. We polluted the breeze and the air. Very few have acted with a consciousness of transformation wanting to change the system. That chaos is what today governs us. Until now, we were playing with fire. We put ourselves off balance. And then, a virus, the smallest of the elementals, the most insignificant creature before the eyes of the younger brothers, forced us to stop the pursuit of the race, without knowing after what we were running. That virus became a great teacher, an authentic messenger.[1]

Paola sees the Pivot Projects as another journey of discovery. She hopes the organization influences policymakers and people in cities and communities, but she believes it should count as a success even if the only minds it changes are those of its participants. "If only the individual is the pivot, the Pivot Projects will have been worthwhile," Paola said.

11

Bright Ideas

In its essence, Pivot Projects was designed to be an ideas factory. The goal was to develop collaborative intelligence by combining the knowledge and creativity of its participants with SparkBeyond's artificial intelligence (AI) machine and a universe of publicly available information. The humans would produce conceptual models of how the world works and, by working with the AI, they would discover ideas about how the world could work better. The group would collaborate with communities to match their needs and ambitions with the best techniques available for helping them to become more sustainable and resilient.

Members of the group, individually and as part of the collaborative process, developed a flood of ideas. Peter Head (a co-leader) and Damian Costello (who led the Economics, Law, and Politics workstream) were veritable idea geysers. But the group also harvested some of the best ideas and innovations that already existed, put its own spin on them, and brought them to the attention of everyone, from national sustainability leaders to community groups in small villages. This chapter spotlights in a digestible form some of the most compelling ideas and solutions that the group developed or championed. I selected for potential impact, novelty, and coolness.

A New Model for Problem Solving

One of the biggest limitations society faces as it confronts climate change is the fact that problem solvers typically operate in relative isolation from one another—with climate scientists on one planet, computer scientists on another, government leaders and planners on others, and each industry on its own whirling sphere. Issues emerging in natural, social, technical, and business systems are viewed and addressed separately. That makes it difficult to share information, collaborate, take on problems that involve multiple human or natural systems, and plan for a better future. Since Peter was a bridge project leader, he has been developing a model for complex problem solving. With each major shift in his career, he added new layers to the model. With Pivot Projects, he added yet others. He calls it integrated systems planning.

In its simplest form, integrated systems planning is the practice of viewing the systems of nature and society holistically and then combining expertise and knowledge from a wide variety of domains to solve large-scale problems. For Peter, the big conceptual breakthrough came in 2004 when he was invited to join Arup Group, the global engineering company. His task was to create its first integrated planning practice. Previously, Arup's planning consultancies—urban design, land use, master planning, environmental planning, economics, and policy—operated separately. He melded them together so they could help the firm's clients achieve sustainability goals no matter what kind of project they were working on. The most significant projects Arup undertook in this way were the Ecological Cities in China, where the Chinese government set out to build more sustainable cities from scratch.

In 2020, Peter had been invited to help prepare policy recommendations concerning sustainable infrastructure for the Group of Twenty (G20), the forum for international cooperation among the leading nineteen countries and the European Union. The 2020 summit had been scheduled for November 21–22 in Saudi Arabia, but it was conducted virtually because of COVID-19. Peter and a handful of colleagues authored a policy brief calling for the G20 to encourage the use of integrated systems planning, including systems modeling, data analytics, and sustainable procurement practices. The authors cited the World Bank estimate that countries could reduce the cost of achieving the UN's Sustainable Development Goals by 40 percent if they adopted this kind of approach. "Most infrastructure projects are very narrowly focused, and they're missing opportunities," said Rowan Palmer, one of the coauthors of the brief who has worked as a researcher for the UN Environmental Program. "If you think more broadly, you can achieve triple-win outcomes—with better infrastructure, better sustainability, and you save money in the long term."

Pivot Projects itself was an example of integrated systems planning. It combined the elements of a diversity of thought, multidisciplinary collaboration, systems thinking and modeling, and AI, and applied them to both high-level policymaking and street-level problem solving. It seems possible that Pivot Projects' model for getting things done might turn out to be one of its most important contributions if others take lessons from it.

Integrated systems planning may seem like a no-brainer, but, in spite of the obvious benefits, it is not in fact practiced widely by governments, professions, or industries. Peter hoped that the policy brief[1] he coauthored would be adopted by the G20 and would also influence the twenty-sixth United Nations Climate Change Conference of the Parties (COP26) climate meeting that was planned for 2021. He's an optimist. "Eventually we won't call it integrated systems planning anymore. We'll just call it planning. It will be the way planning is done," Peter said. "The same will go for sustainable development. At some point all development will be sustainable."

Transforming Industries and Domains

You can see how the integrated systems planning approach could be applied to a wide range of global problems and industries. The COVID-19 pandemic is an obvious example. Some epidemiologists and public health experts from around the world did collaborate to coordinate responses to the pandemic and to share data about the behavior of the virus. The United Nations and the World Health Organization sponsored webinars and other programs aimed at sharing knowledge. But the global response overall to the spread of the virus was notable mainly for the weakness of cooperation and coordination across governments and societal domains. In fact, the Trump administration actually withheld vital data from researchers. It was an every-man-for-himself response—and it was a disaster.

In the wake of the pandemic, you can envision how entire industries or clusters of industries could have come together to confront their shared challenges using an integrated systems planning approach. The travel and tourism industries were hit particularly hard—so hard, in fact, that it seemed likely that they would have to make substantial changes in their structures, business models, and operations if they hoped to regain the vitality that they had enjoyed in 2019. The World Travel and Tourism Council (WTTC) did convene in the midst of the crisis to develop a global plan to save the industry, but it focused on securing government bailouts rather than on making fundamental changes that would set it on a more sustainable course. That omission came in spite of the fact that greenhouse gas emissions from commercial aviation are on track to triple by 2050 and account for 25 percent of the global carbon budget.[2] This industry had its head firmly planted in the sand—to the detriment of Earth's 8 billion human residents, a large percentage of whom have never traveled in an airplane.

Industries can do better. In its early months, Pivot Projects launched an industry initiative, the Built Environment Pivot Institute (BEPI), which had the potential to realign the construction and development industries and make them more sustainable. Again, this came from the fertile brain of Peter Head. He had risen to the top echelons in the bridge-building field and understood the interlocking relationships of construction, environmental protection, transportation, and economic development. To get the institute off the ground, he recruited Kevin Bygate, a longtime steel industry executive. The goal was to create a global membership alliance of individuals and organizations associated with the construction industry, including engineers, architects, planners, investors, developers, government leaders, product and material suppliers, and building owners. Members would pledge to help achieve zero-carbon targets by 2050. They would adopt the integrated systems planning approach. Like Pivot Projects, BEPI would bring together a wide range of people and organizations to address shared problems and would use systems modeling and AI.

The idea was that this organization would complement the Leadership in Energy and Environmental Design (LEED) certification program for buildings developed by the U.S. Green Building Council. Each of the professions and subindustries has its own standards. BEPI would build on them and coordinate between them. Professionals who followed the BEPI code would be able to add a P (for "Pivot") to their credentials. "This is a transformative approach," said Kevin. "As professionals, we have the responsibility to do this ourselves. We can't wait for government or boards of directors to lead. Practicing professionals have a moral and ethical responsibility to address this now."

Living by Principles

From the earliest days of Pivot Projects, its leaders recognized the importance of adopting a set of guiding principles. Their goal was clear from the start: to help set the world on a path to becoming more sustainable and resilient. To flesh out the details, they adopted the UN's Sustainable Development Goals (SDGs), which had been agreed to by representatives from the UN's 193 sovereign states through a multiyear process of consensus building. For Pivot Projects' leaders, it was also critically important for the group to adopt a set of principles that would help shape the discussions, the systems models they built, the questions they asked of the AI machine, and their interactions with people in communities. They had sought to attract a diverse assortment of participants, but white males still predominated. The principles had to help correct for that diversity deficit, and, for that reason, they couldn't be fashioned solely by members of the group. That would be too narrow.

To be credible, Pivot Projects had to find a set of principles that were universal, or as universal as possible in a world where people hold many often-conflicting views and values.

The best option they had was the Earth Charter, a global declaration of commitment to holding humans and the planet in balance. (An outline of the charter is appended to the end of this chapter.) The process of developing the charter was launched in 1994 by Mikhail Gorbachev, former premier of the former Soviet Union, and Maurice Strong, who had led the first UN Earth Summit in 1992, which was a precursor to the Paris Agreement on climate change. The charter's sixteen principles were developed by the Earth Charter Commission over a period of six years through consultation with people from around the world. The charter, finalized in 2000, argues that environmental protection, human rights, equitable human development, and peace are interdependent and indivisible.

The Earth Charter came to Pivot Projects via a circuitous route. One of the early commissioners was Rabbi Awraham Soetendorp, a friend of Jeff Newman. In 2005, Awraham had invited Jeff to help popularize the Earth Charter in the United Kingdom, where it hadn't had much impact at the time. Jeff introduced the charter to his friend Peter Head. When Pivot Projects launched, Jeff signed on as a co-leader of the Faiths workstream. He invited Awraham to join. (In chapter 3, remember how Awraham delivered a benediction of sorts in one of the first Zoom gatherings.) In a matter of weeks, the group adopted the charter as the basis of its core principles and as a beacon, a statement of faith, and an operating guide. Jeff explained, "What really excites me about the Earth Charter was always the holistic, systemic approach that it totally enshrines—the sense of wholeness, the recognition of oneness, of unity, of our shared human potential."

The Pivot Projects leaders urged any group or organization pursuing a social cause to adopt guiding principles. Any effort to transform the way systems work must take into consideration the needs, rights, and values of a large and diverse set of people. At the same time, principles can't just sit on a shelf; they have to be acted on. "Pivot Projects will be judged, rightly, not on what it says but what it does," says Jeff.

A Modest Proposal Regarding the Global Economy

Modern capitalism is a remarkable phenomenon. It has powered tremendous advances in science, knowledge, and industry. It has built megacities and megafortunes. And it has helped raise the standards of living and the aspirations of billions of people. Yet the gap between rich and poor is widening rapidly; the system has been built on exploitation of labor and overconsumption of goods and resources; and, yeah, that elephant in the room: the way capitalism is being practiced today has set the world on a collision course with climate catastrophe. It was the job of the Economics, Law, and Politics workstream to sort through this stuff and develop a solution.

The first impulse was to burn the edifice down. In fact, some of Damian Costello's early slide shows featured an illustration of a house that was engulfed in flames. He had radical impulses, in spite of the fact that he worked as a consultant for giant corporations. It seemed that the only way to avoid calamity was to identify the alternative economic model with the best chance of delivering a rapid transformation and get behind it. Gradually, though, his view evolved, and he took the rest of the group with him. They ended up with a new thought, which Damian called the cellular economy.

Damian Costello

The cellular economy idea was that, within the existing economic and social systems, groups (or cells) of people linked by geography or shared beliefs and aspirations would coalesce and launch initiatives aimed at improving life locally or globally, making it more sustainable, peaceful, and just. The group was inspired by economist E. F. Schumacher's mantra, "Small is beautiful" and also by the global social entrepreneurship movement. These cells might be mission-driven companies, social enterprises, villages, or neighborhoods. They would be focused more on fostering shared values than on profits or on getting ahead in a traditional individualistic or nationalistic sense. They would operate independently of one another, but they could be linked by communications networks, and they could affordably tap into powerful data analytics and AI resources via cloud computing services. The cells would perform social, technological, and economic experiments aimed at solving real-world problems. "You have a million experiments," Damian said. "When enough of these small successes emerge, you no longer have a social pyramid like the one that underpins today's rigid system. The emerging system will be flat; self-organizing; dynamic; and constantly moving, innovating, and evolving."

The group felt this socioeconomic pivot was consistent with the great sweep of human history. Early feudal systems were patriarchal, hierarchal, and oppressive to the masses. Individuals were subservient to the lords of church, manor, and military. With the Reformation, the Renaissance, and the Enlightenment in Europe, individuals gradually gained independence (although in much of the rest of the world, slaveholding and colonialism were continued forms of feudalism). Large numbers of individuals were blessed with independence and freedoms, yet they were not truly free. They were and are still dominated by the corporate lords of the economy and the idea that big business is better at organizing society than are citizens through democratic structures. Meanwhile, greed and consumerism, the hyped-up horsemen of the modern apocalypse, are laying waste to nature.

Pivot Projects aligned itself with a number of alternative economic theories and models that are designed to make the global economy more sustainable and to distribute wealth and resources more equitably.

These include the concept of the circular economy, which aims to reduce resource consumption and waste through recycling and reuse; regenerative economics, which focuses on using natural assets, including wind, sun, and tides, to help society operate more sustainably; and Doughnut Economics, which measures efforts to provide equitable access to life's essentials while helping society avoid overconsumption. None of these theories represents a wholesale shift to a new economic system. They're all amendments to capitalism.

The same was true for the Economics, Law, and Politics workstream. While its initial goal was to try to identify or develop a singular vision for a new global economy, the group instead discussed a cluster of ideas that might help reform capitalism but not replace it, at least not immediately. The cellular economy was the most fully imagined of the three ideas. The Pivot Projects people believed that the next major paradigm shift for the global economy is will be too diverse and organic and thus cannot be foreseen and categorized in traditional monolithic terms. The Pivot Projects people envisioned a hybrid model emerging that has centralized power structures coexisting with an array of cellular alternatives that they hoped would gradually come to predominate.

What would be the role of Pivot Projects in this economic pivot? Not what you might think. Damian and his crew didn't see themselves or Pivot Projects as saviors. The group, they thought, would be one of many cells committed to sustainable innovation. Indeed, there are already a large number of groups around the world doing this kind of work, so Pivot Projects' role would be to help the movement increase its membership and accelerate. "Our job is to midwife the birth of the new organism," said Damian. "We don't need to lead the charge. We can help people get started and trust them to change the world in their own way."

Facing Facts: There *Are* Limits to Growth

In 1972, the Club of Rome published a multiyear study by researchers at the Massachusetts Institute of Technology (MIT) called *The Limits to Growth*.[3] I cited this work earlier in the book as one of Colin

Harrison's first exposures to systems thinking. In fact, it was one of the first research projects that used systems thinking as its basis and systems modeling as its core methodology. The MIT researchers concluded that the population growth rate at the time would rapidly overtake the planet's resources. Unless society switched gears, the limits to growth on Earth would become evident within 100 years, leading to a sudden and uncontrollable decline in population and industrial capacity. A follow-up study published in 2004 showed that subsequent population growth and resource depletion had tracked the original projections closely.

In the midst of the COVID-19 pandemic, with global population heading toward 8 billion and with environmental destruction and social dysfunction raging all around, Pivot Projects' Sustainable Infrastructure workstream performed some basic calculations and concluded that, not only are there limits to population growth, but there were also limits to the ability of economic growth to help solve the problem. "The way we operate today depletes resources and distributes benefits unevenly," said Stephen Hinton, who led the workstream.

At the time, there were a number of large-scale green infrastructure proposals making the rounds of legislative bodies globally. They carried huge price tags. One study, *The 2018 Report of the Global Commission on the Economy and Climate*, estimated that a global investment of $90 trillion was needed and called for investments of between $5 and $7 trillion per year for fifteen years.[4] While China and the European Union had committed to making massive green infrastructure investments, other countries, most notably the United States, had recoiled at the costs of achieving Earth-scale sustainable infrastructure. These countries felt that they couldn't afford to make the switch. (Ironically, the public and private sectors were making huge investments in maintaining or purchasing existing, and unsustainable, infrastructure.) Just as troubling, in spite of earnest efforts by some countries to become more sustainable, not a single one was meeting the basic needs for its citizens at a globally sustainable level of resource use.[5]

The Sustainable Infrastructure workstream recognized a conundrum. It seemed that robust global economic growth (as measured by global world product [GWP], a combination of all nations' gross domestic products [GDPs]) would be necessary to help pay for improvements in sustainability, yet economic growth itself was at the core of the sustainability problem. True, the global growth spurt of the early twenty-first century had helped bring many millions of people out of abject poverty, yet the economic divide between the richest humans and the rest had become wider.[6] At the same time, it was clear that the world economy could not continue to grow at rapid pre-COVID-19 rates without turning the planet to a cinder.

A way around this problem, the group saw, was to reduce overall consumption substantially. Stephen Hinton said that it is critical for society to shift from a focus on GDP growth and achieving a middle-class lifestyle to a focus instead on living sustainably and improving living conditions for the many. This requires sacrifice by some for the benefit of all. Stephen is a mix of pessimism and optimism. One day in a Zoom meeting, he delivered this soliloquy:

> Pivot Projects should be very clear about, look, we can't sustainably provide everyone with a middle-class lifestyle on this planet. It's impossible. And we can't even sustainably support those who already have a middle-class lifestyle in the same way. Give it up. But there is something we can do: Everyone can be fed, housed, have a job, water, safety, security, and a community. It just won't look like most people's middle-class dream.

But Is a World with Flat Growth Fair, and Is It Doable?

The question of fairness came up repeatedly in Pivot Projects meetings. Was the group biased because of its demographic makeup? If economic growth was toxic to the planet and ultimately to humans, how could resources and wealth be distributed in ways that were fair and equitable? Practically everything society does involves trade-offs. When an action is taken or a new policy is put in place, typically, somebody wins and somebody else loses. The SparkBeyond AI might help level the playing field. One of its key capabilities was its ability to help people evaluate conflicts and resolve trade-offs. But the weight and responsibility for navigating fairness falls squarely on the shoulders of humans, with all of our prejudices, blind spots, self-interest, and ignorance.

No question of fairness was more central to the Pivot Projects team than the one that Stephen Hinton had raised. It appears that rapid top-line global economic growth is not sustainable. As a species, we must consume less. The wealthy can't continue to live the way they do, nor can the middle-class. And, as Stephen said, we can't promise poor people a traditional middle-class lifestyle. How do you convince people of the fairness of this conclusion? And how do you persuade them to live with less consumption and lesser material dreams?

Stephen saw a trade-off worth making. By focusing on ending suffering and using resources in a more sustainable way, society would lay the foundations for a more stable, just, and peaceful world.

But others in the group were unsettled by the idea that, because of the need to combat climate change, people in developing nations might be denied the material benefits that others elsewhere already enjoy. Angela Fernandez, an American whose family came from the Dominican Republic, said that she hoped that poor people and people in developing nations won't be shortchanged. "If you stunt economic growth

Angela Fernandez

worldwide, the developing countries will remain developing for a very long time," she said. "That's unfair. Everybody wants to improve economically." She called for massive, low-cost loans to help transfer key, sustainably oriented technologies such as desalination to developing countries.

On the other hand, was it realistic to expect that rich or middle-class people would be satisfied to have and consume less? Damian Costello was highly skeptical. "I don't see anybody willing to make sacrifices. Even COVID-19 has barely moved the needle. We in the West are not facing up to any level of sacrifice that people in developing nations would consider fair."

These were hard questions. Pivot Projects people became animated and even exasperated during discussions about them. At one point, Shulamit Morris-Evans, the young UK Extinction Rebellion activist, was frustrated when the others in the group seemed unwilling to envision a world where the economy would not focus on returns on capital investment but instead on meeting the needs of the people. "What's problematic is this social construct we have invested in to such a degree that even though it's killing us we can't think outside of it," she said. "Why can't we say *these* are the resources we have and *these* are the needs, and then match them up, and live sustainably?"

Others in Pivot Projects believed that the economic tensions were resolvable. Peter Head cited the example of Japan, where GDP growth had been about zero since the 1980s, the country recycled and ran energy-efficient transportation systems, and yet the population enjoyed a high standard of living. He also pointed to scientific studies concluding that it was possible to live a fulfilling life while using much less energy.[7] Ian Richardson, from Northumberland in the United Kingdom, matched Peter's optimism: "We can preserve the lifestyle in the West by moving to better technologies and renewables. We don't have to turn the lights off." Damian hoped that the cellular economy, if it evolved as he imagined, would make it possible to satisfy Maslow's hierarchy of needs for humanity without crashing head-on into planetary boundaries.

In early 2021, there was little evidence that the world was going to shift away from the fundamentals of capitalism, but at the same time, there were glimmers of hope concerning sustainability. The more progressive energy companies were moving to renewables; governments were more willing to invest in sustainable infrastructure; climate change deniers had been thrown out of office in the United States; and that idea of fostering a cellular economy seemed promising, at least in one little corner of the universe.

Teaching Kids How to Live Sustainably

If sustainability and fairness are to be achieved, it will likely require a massive shift in the attitudes and lifestyles of individuals. A couple of billion people in the world, the "haves," will need to figure out how they can want less, consume less, and waste less. Meanwhile, poor people might have to adjust their aspirations vis-à-vis material wealth. This will require a radical change in education.

Tremendous progress has been made over the past two decades in providing universal primary education in developing nations, yet in many countries, rich and poor, education systems are failing children and society. Children, especially those in poor neighborhoods, aren't being taught the subject matter and the skills that they will need to succeed in the twenty-first century. That situation was a given in the discussions conducted by the Pivot Projects Education workstream, but rather than trying to fix the whole thing, they focused on one critical area: teaching about nature and the environment. On the personal level, living sustainably can't be forced by government fiat. People have to understand and accept the need to change.

In most countries around the world, environmental sustainability is not part of the primary school curriculum, and the Education group believed that rectifying that situation would be key to dealing with climate change in the decades ahead. Rather than pressing to add traditional lessons to already overwhelmed teachers and students, they explored ways of moving education about the environment and sustainability out of the classroom. This was one of the most diverse of the workstreams, including not just reformers from Northern Europe and the United States but young people from Syria, Vietnam, Indonesia, and

Bangladesh. Floris Koot, an educator from the Netherlands, was the most forceful in his critique of the status quo: "We are training kids to be part of the murdering machine that kills the planet. It's hard to step out of it, but we've got to change that."

Floris had been a cofounder of Knowmads Netherlands, a tiny experimental business school focused on helping young adults become leaders and innovators in the sustainability space. Students there learned in part by dreaming up and launching social entrepreneurship projects. When they began, they were asked to answer a handful of questions: (1) Who are you? (2) In what world do you want to live? (3) How will you help create that world? (4) What business project will help you do that while making an income? (5) How will you put that into the world? The school was something like Silicon Valley for sustainability. Unfortunately, it failed during the COVID-19 crisis.

The Education workstream focused on approaches that could conceivably be implemented on a mass scale globally. Its members searched for promising examples of teaching environmentalism by taking kids out of the classroom. They spoke to teachers and administrators about what could be done. This had to happen at the grassroots level. The idea was that, when Pivot Projects engaged with communities, they'd be able to help launch innovative environmental education programs there. One of the most exciting projects they spotted was the so-called Wild Mile, a mile-long floating ecopark on the sides of a steel-lined river near downtown Chicago. A nonprofit called Urban Rivers had teamed with the city to build the park with the goal of rewilding the river, promoting outdoor recreation, and teaching schoolchildren about nature and ecology. Said Alan Dean, leader of the workstream, "We have to get the message down to the young, and they will teach their parents."

Using Regulation as a Tool for Transformation

One of the most powerful forces at work in society today is the digitization of marketplaces. Leading examples are online marketplaces such as Amazon, Uber, and Airbnb; electronic equities and commodities exchanges; advertising platforms including Google and Facebook; and cloud computing services such as Amazon Web Services and Microsoft's Azure. They provide tremendous efficiency and convenience for sellers and buyers, and they are demolishing traditional business models and industries. This rapid-fire transition concentrates power and wealth in the hands of the owners of the most successful marketplaces (that is, technology platforms), and it eliminates jobs and devalues labor.

This transition was of particular interest to members of Pivot Projects' Technology workstream, which was led by Rick Robinson. While they applauded the benefits that these marketplaces bring to society and the economy, they were concerned that the sheer power of the platform owners was causing an imbalance that would have significant negative effects on society and the environment over the long term. For instance, by operating a global digital ride-sharing service, Uber catalyzes a shift from energy-efficient mass transportation to energy-guzzling private vehicles. At the same time, because of its market dominance, Uber essentially controls pricing for an entire industry. In the future, when it shifts to autonomous vehicles, it will eliminate millions of jobs worldwide. Like other businesses of its type, Uber's purpose is to maximize profits while driving within the lines of the law. The ultimate effect is to exaggerate the inequities that already exist in society. "The rich are getting richer, and the poor are getting poorer," Rick said.

The antidote, Rick and his colleagues concluded, is intervention by governments using innovative regulations and tax policies to minimize the negative effects of the rapid shift to digital marketplaces. Rick's thinking was influenced by that of American economist Elinor Ostrom, who won the Nobel Prize in Economics in 2009 for her work on the governance of the commons. The commons are the cultural and natural resources that we share, including clean air and water and well-functioning ecosystems. Ostrom's work draws heavily on systems thinking and complexity theory. She laid out design principles for managing and protecting the commons.

Ostrom's design principles could form the basis for the regulatory innovations, Rick thought. We have already seen proposals along these lines. For instance, carbon taxes are designed to protect the commons by increasing the price of fossil fuels and decreasing demand for them. Rick and his colleagues proposed the creation of new regulatory regimes designed to reward capitalists and corporations that take actions that help society achieve the UN's SDGs and penalizing those who make it more difficult to achieve them. "Most corporations are focused solely on maximizing profits," Rick said. "Unless the law says sustainability matters, it won't."

Rediscovering the Fifteen-Minute City

Through most of human history, people lived in villages, towns, or neighborhoods where they could walk from their homes to work, play, pray, experience nature, shop, and, if they were fortunate, enjoy art, theater, and music. It wasn't until the twentieth century that places got so big that people couldn't conveniently walk across their towns or cities and it wasn't until the second half of the twentieth century that, in many cases, people's homes, workplaces, shops, and other places of interest were miles and even hours apart and required cars, buses, and trains to make the trip. We burned a lot of fossil fuels getting from one place to another, and, by fragmenting the elements of our lives in this way, we made it difficult to maintain relationships, shared interests, and empathy for one another. Thus were born some of the ills of the early twenty-first century.

The fifteen-minute city is a movement to regain some of the physical and social features of the villages, towns, and neighborhoods of yore. The idea is that, if we create places where people live, work, and play within a fifteen-minute walk or bike ride, we can live more sustainably and can regain some of the benefits and satisfactions that come from living in a tightknit community where people know and care about each other. The concept and catchphrase were developed by Professor Carlos Moreno of the Sorbonne in Paris and were popularized by Paris Mayor Anna Maria Hidalgo. After the COVID-19 pandemic struck and people were isolated at home, the idea spread globally. It was championed by the C40 Cities, a network of progressive cities around the globe as part of their Green and Just recovery plan.[8]

Aneta Popiel, the leader of Pivot Projects Built Environment workstream, fell in love with the concept when she first heard about it. A young Polish woman living in London, Aneta works as a manager for workplace-as-a-service for a large building construction and management company. Among other responsibilities, she has managed a coworking space, organized arts events, and run property innovation hackathons. As wonderful as London is in many ways, it's also hard to get around and it is a fragmented city. Aneta has experienced a human-scale city on her frequent visits to Tel Aviv, Israel: a bustling, vibrant place full of experimentation and camaraderie. "The fifteen-minute city is a more livable city," Aneta said. "It's really about building community. Great things can happen when you put people together. They share their skills and you give them more say over what happens in their neighborhood."

Aneta believes that Pivot Projects should offer expertise about the fifteen-minute city concept when it engages with cities and helps local people identify approaches to redevelopment that might enhance community life. Aneta noted that people often think about sustainability in negative terms: as something that requires them to give up things that they enjoy. It doesn't have to be that way. "Instead of a stick we need to have a carrot. The fifteen-minute city is the carrot. It allows people to live their best life without destroying nature," Aneta said.

Making Art for Resilient Communities

When my wife, Lisa Hamm, got her certificate to teach art in the public schools in the 1980s, her first job had her pushing a cart loaded with art supplies from classroom to classroom in a middle school to keep the kids occupied while the regular teachers had their breaks. That's how unimportant art education was in a town in Connecticut, one of the richest states in the union. In many places, sadly, little has changed since then.

But in other places, in some of the most vibrant cities and towns in the world, art has been recognized as being essential to the vitality of local economies, to the appeal of living in and visiting places, and to the well-being of people who live there. In the time of COVID-19, it has become manifestly clear that art of all kinds—and especially public art—is critical to the resilience of communities. From around the globe came reports of communities responding to the pandemic and other social stresses with public art projects. To help children in the United Kingdom deal with the ravages of COVID-19, Arts Council England invited young people to submit art and writing depicting their responses to the pandemic and featured the work on the Google Arts and Culture platform. In dozens of U.S. cities, Black Lives Matter (BLM) activists were encouraged to paint their slogans on city streets in huge block letters.

One of the more compelling arts projects emerged in Mexico's San Miguel de Allende, a city already famous for its many street murals. City leaders there were concerned that young people were not wearing masks when out in public. They launched a two-pronged effort to raise awareness and change behavior: putting masks on historic outdoor statues and commissioning young people to paint murals depicting famous paintings with their subjects wearing masks. Ten such murals were completed in a single day, including riffs on Vermeer's *Girl with the Pearl Earring* and a van Gogh self-portrait. It worked. Young people began wearing masks more frequently, and the city had a low infection rate. "When it comes to something like a pandemic, there are a lot of emotions and mental health issues," said Colleen Sorenson, an American expat who helped foster the mural project in San Miguel de Allende. "Art is necessary for all-around well-being, and young people must be given a voice. Let them speak on the street."

Even before the COVID-19 pandemic, public murals were spreading rapidly around the globe. And they're not just in big cities. Murals have proliferated in modest cities and towns where tourists rarely tread, but where local people see great value in using art in public spaces to build a shared identity. These public art projects are elements of a global social movement called cultural placemaking that emerged in the 1960s in New York City. It was heavily influenced by the urban design soothsayer

Klimt-inspired mural by Jesús (Juice) Valenzuela

Robert J. Hawkins

Jane Jacobs, who led a movement to revive Manhattan's West Village. The idea is that city officials, community groups, or even tribes of social misfits can reshape public spaces by making and displaying art in them.

The Pivot Projects Arts and Culture workstream believed that the arts have a critical role to play in a pivot to a better society. Arts of all kinds help individuals, communities, and the broader society see things in new ways and help to create conversations that have the potential to build bridges between people with different backgrounds and points of view. "Art is not a social lubricant to help communities feel better about themselves and their surroundings. It speaks directly to the source of our humanity, and, in so doing, inspires and informs even the most frozen of hearts," said Damian Costello, who dropped in on the Arts and Culture workstream. "It is one of our most powerful weapons in the war against dehumanization. This is the essence of the role of art in society."

People in the workstream saw the value in promoting public art, so they looked for ways to encourage the creation and sharing of art related to COVID-19 and the climate crisis. One such effort was virtual: they set up a channel on the PivotProjects.org website where they featured arts of all kinds that were created during the COVID-19 pandemic—everything from murals to tattoos, to TikTok videos. In addition, they planned to offer to help communities foster public art as an element of their campaigns to become more sustainable and resilient.

Regenerating Coastal Forests to Save the Planet

It is well established that forests play a critical role in slowing climate change. Approximately one-third of the CO_2 released from burning fossil fuels is absorbed by forests every year. Halting the loss and degradation of forests and promoting their restoration have the potential to contribute to over one-third of the total climate change mitigation that is required by 2030 to meet the objectives of the Paris Agreement.[9] Massive efforts are underway worldwide to save and regenerate forests. Yet there's another contribution by forests to climate stabilization that has long gone unrecognized: the role of coastal forests in addressing drought.

The role of coastal forests was the theme of the section coauthored by Peter Head in a UN special report on drought, which was published in 2021. "The role of forests in much of the world's freshwater cycles needs to be properly recognized," said Peter. "Climate scientists focus on the role of climate change in drought, but forests play a major role, too. We need to use reforestation and regenerative agriculture to deal with drought."

Before the rise of the Roman Empire, the Mediterranean Sea was ringed by forests. To support a population that at one point reached nearly 60 million, the Romans cleared massive expanses of land for agriculture and used lumber wherever they went for shipbuilding, housing, mining, and other activities. Much of the land around the Mediterranean Sea was denuded. The result: desertification.

Think of forests as huge biotic pumps. They suck water from the earth and the air, and they recycle it back into the atmosphere. If you remove trees from coastal areas, the environment loses the dynamic interplay between the moisture over the sea and the moisture in the forest. Rainfall diminishes, drought occurs, trees die. The biotic pump sputters and finally conks out. We see this phenomenon not only around the Mediterranean but on the coasts of Brazil and Australia.

In their UN report, Peter and his coauthors cite examples from Ethiopia of the regeneration of forests having a positive impact on rainfall. In the Tigray Region, for instance, communities combined regeneration of forests with soil and water conservation measures to transform the area into one of the most water-secure areas in all of Ethiopia.[10] These small-scale successes hint at what could be accomplished on a global scale.

Peter tipped his hat to the Trillion Tree Campaign, which was announced at the 2020 World Economic Forum meeting in Davos, Switzerland. The goal is to plant 1 trillion trees worldwide by the end of the decade. But he also challenged the campaign's logic. "Just planting a lot of trees isn't necessarily the answer. What really matters is where you plant them," Peter said. "It has to be part of a systems solution—related to the winds, seas, and mountains. You're wasting your money if you do it the wrong way."

Dealing with Wildfires the Indigenous Way

A couple of days before I interviewed him, Chris Medary, a member of the Climate Risks workstream, had been rock climbing in the Colorado mountains with his brother when the sky suddenly turned orange and ashes started falling around them like snow. It meant a wildfire was approaching. Fortunately, they got clear of it. "I was really scared," said Chris.

He's also really determined to do something about it. Chris was pursuing a master's degree in environmental management at Western Colorado University. One of his passions is studying how indigenous people interact with the natural environment. In the Gunnison River valley, where he lives, the Ute people traditionally thinned out forests by carefully burning underbrush so the forest floor was clear. That's now called prescribed burning. They did it to make it easier to hunt game, but it had another positive side effect: it reduced the likelihood of large, ruinous forest fires. Chris believes that modern-day humans have a lot to learn from indigenous cultures. They were intuitive systems thinkers and understood how they fit in. They didn't fight nature. "Living in harmony with nature makes a culture more resilient," he said.

The summer of 2020 was the worst on record for wildfires in the American West, and it had forced state and federal land managers to reassess their approach. Current practice was to suppress all fires on lands controlled by the Department of the Interior on behalf of the corporations that harvest lumber there. But this practice wasn't working, and it was extremely expensive. Now, conservation groups were urging them to use the ancient Native American burning techniques on a large scale. It was unclear if they would listen.

Chris's study of forests as systems melded his university studies with his volunteer work for Pivot Projects. He was pitching in with the Climate Risks workstream, which was working with local groups to create a practical problem-solving framework to help municipalities and regions evaluate and address climate risks. "We need a pivot in thinking," Chris said. "We're not separate from the land. It's not something to control or suppress; it's something to be actively engaged with."

Building Self-Powered Villages

In the developing world, a mass migration is going on as millions of people flee rural villages annually for megacities such as Delhi, São Paulo, Mexico City, Cairo, Dhaka, and Lagos. They are being driven by poverty, climate change, competitive pressures on small-holder farms, and the desire for opportunities for themselves and their children. For many, however, the miseries that await them in cities dwarf those they fled in the countryside. What if rural life could be made more prosperous and sustainable so people would stay home? That's the idea behind SUNRISE, a global initiative led by Swansea University in the United Kingdom aimed at helping communities generate their own solar power and use it to save money and kindle economic growth.

At the core of the initiative is the idea of turning buildings into power stations capable of producing all of the energy they need and making excess energy available for other uses. The team is using an experimental material, perovskite solar cells, which can be applied to structural steel as a coating. The material is cheap to produce, is simple to manufacture, and requires minimal maintenance. The extra energy produced by such buildings has the potential to transform village life by providing power for farming, small industries, and Wi-Fi networks. That could help villages stem the tide of outmigration.

Peter Head and Ian Mabbett, an associate professor at Swansea University who is chief operating officer for SUNRISE and a member of Pivot Projects, were instrumental in launching the initiative. The first buildings are being constructed in villages in India's Maharashtra state through an alliance with Tata Group, the country's largest business. Peter and Ian hope projects in Mexico, South Africa, and Kazakhstan will soon follow. They plan to introduce the model to communities around the world through Pivot Projects. "We can bring all sorts of services to rural places, including sanitation and water purification," Ian said. "We'll engage with the communities to see what's important to them."

A Personal Path to Recovery

In a number of cases, participants in Pivot Projects brought ideas that were intensely personal. One example was Bill McKenna, an American living in Austin, Texas. A few years earlier, he had received a PhD in science, technology, engineering, and math (STEM) education from the University of Texas at Austin. During the pandemic, Bill had patched together a living by consulting for education research projects and assisting a contractor friend in building houses and remodeling. But his avocation was recovery. He had struggled with alcohol use during his graduate student days and had been sober for over eight years. He participated in Alcoholics Anonymous programs. A big focus of his was helping other people restore balance in their lives. He and a friend had set up a website where they posted audio interviews with people in or near recovery in order to demystify addiction recovery and show how it applies to everyone.

Bill saw strong parallels between substance abuse disorder and addiction to consumption. While addictions to alcohol and drugs destroy peoples' lives and those of their loved ones, it seemed to him that addictions to overconsumption of material things have set humanity on the path to planetary destruction. He believed lessons learned from the substance abuse recovery journey could help guide society's recovery from overconsumption and from blindness to the perils created by abuse of the environment.

Many studies have shown that drug and alcohol abuse are related to traumas and frustrations that people experience in their lives. Many people self-medicate in order to anesthetize that pain. Bill believed that the same is true for societies. People fill unmet needs for love, connection with others, purpose, and self-worth by accumulating clothing and vehicles, by living in overlarge houses, and by consuming overlarge meals. As a result, we ignore truths about ourselves and abuse the natural environment. We may not mean to do so, but we do. "Most forms of overconsumption are not fatal to the individual, but, taken in aggregate, they may be fatal to the world and maybe to society," Bill said. "The multipliers are powerful. We're sucking the planet dry."

As with recovery from substance abuse, recovery from environmental abuse is a multistep process. The four key steps, Bill believes, are (1) recognize that you have a problem, (2) be honest to yourself about the causes and effects of your problem, (3) throw yourself into the service of others, and (4) stay involved in a community of people with similar commitments. In doing so, Bill believes, how you think and how you live will change. The addiction will no longer be soothing.

Bill openly discussed his own challenges during Pivot Projects workstream meetings, and he expressed his views about the parallels between substance abuse and environmental abuse. He hoped to work with others to develop a Kumu model focused on addiction to overconsumption and recovery from it. He felt it was critical to embed that system in the conceptual model that Pivot Projects was contributing to SparkBeyond's AI platform. Bill also looked forward to joining the Pivot Projects teams that engaged with cities and regions. He wanted to discuss the lessons of recovery with people in those places.

Addendum

The Earth Charter[11]

(Outline)

Preamble

We stand at a critical moment in Earth's history, a time when humanity must choose its future. As the world becomes increasingly interdependent and fragile, the future at once holds great peril and great promise. To move forward, we must recognize that, in the midst of a magnificent diversity of cultures and life forms, we are one human family and one Earth community with a common destiny. We must join together to bring forth a sustainable global society founded on respect for nature, universal human rights, economic justice, and a culture of peace. Toward this end, it is imperative that we, the peoples of Earth, declare our responsibility to one another, to the greater community of life, and to future generations.

Principles

I. Respect and Care for the Community of Life

1. Respect Earth and life in all its diversity.
2. Care for the community of life with understanding, compassion, and love.
3. Build democratic societies that are just, participatory, sustainable, and peaceful.
4. Secure Earth's bounty and beauty for present and future generations.

II. Ecological Integrity

5. Protect and restore the integrity of Earth's ecological systems, with special concern for biological diversity and the natural processes that sustain life.

6. Prevent harm as the best method of environmental protection and, when knowledge is limited, apply a precautionary approach.
7. Adopt patterns of production, consumption, and reproduction that safeguard Earth's regenerative capacities, human rights, and community well-being.
8. Advance the study of ecological sustainability and promote the open exchange and wide application of the knowledge acquired.

III. Social and Economic Justice

9. Eradicate poverty as an ethical, social, and environmental imperative.
10. Ensure that economic activities and institutions at all levels promote human development in an equitable and sustainable manner.
11. Affirm gender equality and equity as prerequisites to sustainable development and ensure universal access to education, healthcare, and economic opportunity.
12. Uphold the right of all, without discrimination, to a natural and social environment supportive of human dignity, bodily health, and spiritual well-being, with special attention to the rights of indigenous peoples and minorities.

IV. Democracy, Nonviolence, and Peace

13. Strengthen democratic institutions at all levels, and provide transparency and accountability in governance, inclusive participation in decision making, and access to justice.
14. Integrate into formal education and life-long learning the knowledge, values, and skills needed for a sustainable way of life.
15. Treat all living beings with respect and consideration.
16. Promote a culture of tolerance, nonviolence, and peace.

PROFILE

Ian Mabbett

Welsh university professor and inventor

When Ian Mabbett was growing up in Plymouth, in the southwest of England, he loved to sail. His dad, a machinist, worked in the government shipyard there; at age eleven, Ian got his first sailboat, a ten-foot Mirror dingy, which he took out into the River Tamar and the Plymouth harbor as often as he could. "I loved engineering and science already," Ian said. "But for the first time I could be in control of it. In math class, I studied vectors. When sailing, I could feel the vectors. In my head, I'd be correcting for the tide."

Throughout his life, Ian has been attuned to the practical aspects of science and engineering. He wants his work to have a positive and direct impact on the people around him and also on people in developing nations.

Today, Ian is an associate professor in engineering at Swansea University in Wales. His research has mostly focused on the rapid heating and curing of industrial coatings such as photovoltaics and energy storage materials. He jokingly tells people that he's in the business of watching paint dry, but his domain is actually much sexier than that. His work had played a role in the emergence of a powerful new force in the alternative energy field—the concept of self-powered buildings. In addition to his academic appointment, he's also chief operating officer of SUNRISE, the organization that's behind those solar-powered buildings I described in chapter 11.

At Swansea, Ian has ridden on the coattails of a more senior faculty member, David Worsley. Together, before the SUNRISE project, they developed coatings for protecting structural steel from corrosion. That development evolved into using coatings to attach photovoltaics to steel. Such coatings are less expensive, more durable, and easier to clean than traditional silicon-based solar panels. Working with die-sensitized solar cells embedded in titanium dioxide, the two developed a process for curing the coatings more quickly and inexpensively. They set up a research collaboration called SPECIFIC as a vehicle for testing new green building materials and processes in real-world situations. Some of those experiments produced the foundational technologies for the SUNRISE project.

Many of Ian's obsessions grow from conversations in pubs. He gets to talking to people over a beer, and the next thing is he's off chasing a big idea. In the case of SUNRISE, he had met a project leader from the Bill and Melinda Gates Foundation at a conference. They later connected at a bar. He told the guy his usual line—that he was in the business of watching paint dry. The guy asked him, "Have you ever considered drying shit?" He was referring to the widely used practice of using dried animal dung for fuel in the rural parts of developing nations—which causes health problems. Ian had not, but the conversation set him on a new path: seeking ways to bring sustainable energy production to villages in India. At SUNRISE, they would aim to produce buildings that were not only energy self-sufficient but also would provide power for use in agriculture and other local industries and to improve sanitation and purify water. Buildings would become power stations for villages.

Ian is a connect-the-dots guy. He sees all of the engineering challenges associated with sustainable buildings and construction as part of a constellation of related systems, and he believes buildings won't be sustainable without maximum input in the design stages from people who will live and work in them. "I have this deep inner need to connect everything," Ian said. "I genuinely believe everything I've been working on in my life so far can be connected and the result will be so much stronger than the sum of the parts."

Ian heard about Pivot Projects through Peter Head, who was the chair of the SUNRISE advisory board. When Pivot Projects launched, Peter invited Ian to join. At first, Ian was content to be a quiet observer. He joined the Education, Energy, and Sustainable Infrastructure workstreams, but it wasn't until the Housing and Construction workstream was activated that he became fully engaged. He saw an important role for himself there. "I want to make sure we're not just saying what everybody else is saying. I want to make sure we're challenging conventional wisdom," he said.

He also hoped to play a role when Pivot Projects engaged with cities and communities. He felt empathy for working-class people like those he grew up with in Plymouth. He believed it was essential to reach out and appeal to those who don't believe climate change is a threat or who don't want to make the sacrifices that may be necessary to deal with it. "My ideal world is one of responsible innovation mixed with open and respectful communication," Ian says. "We talk about the difficult-to-decarbonize industries, but no one is talking to the difficult-to-decarbonize people."

12

Connecting

AT 5:04 P.M. on October 17, 1989, a 6.9-magnitude earthquake struck in the mountains south of San Francisco. At the time, I was standing next to my desk in the bowels of the *Mercury News* newspaper offices in San Jose, where I was the technology business editor. When the earth started to heave and buck, I braced myself in a doorway. A bowling pin and a fake "jackalope" wall plaque flew past my head, ceiling tiles fell, and the lights blinked out.

A few minutes later, the entire editorial staff gathered in semidarkness in the main newsroom. The city editor stood on a desk and barked out commands. Then the staff members leaped into action, with many reporters fanning out in cars across the Bay Area and into the Santa Cruz mountains. Some would not return home for days. The Loma Prieta Earthquake killed sixty-three people and injured 3,757 more. It triggered the collapse of a double-decker freeway and a span of the San Francisco Bay Bridge. Yet somehow the *Mercury News* staff members produced that night not just the regular edition of the morning newspaper but also a special earthquake edition. Subscribers recounted living in fear in the blackout overnight but awakening to the reassuring thwack of the newspaper hitting their front steps.

For several days thereafter, the region was roiled by thousands of aftershocks. Some of them were jerky, like a jackhammer, but others felt like we were standing on a ferry moored at a dock, gently but undeniably swaying. The earth was no longer solid under our feet. The earthquake shattered our sense of the impossible. Yet stability and normalcy gradually returned—restored by the competence and dependability of social institutions around us, including the stalwart newspapers of the time and the city, state, and national governments. Ultimately, the Bay Area came back better than ever.

A few days after the earthquake, I took a hike with my family in the Forest of the Nisene Marks in the Santa Cruz mountains, near the quake's epicenter. While a few trees had been felled by the quake, the forest of redwood trees, some of them ancient, was solid and still. Redwood trees have relatively shallow roots, but in a grove, the root systems of multiple trees intertangle so the trees support each other. It's natural resilience. Nature seemed to reassure us that everything would be okay.

Today, faced with the harsh realities of the COVID-19 pandemic, virulent racism, economic inequities, and climate change, a similar positive outcome seems dubious. The institutions and economic and

Redwood grove in the Forest of the Nisene Marks

Gabriela Hamm

social systems upon which we have depended are deeply fissured and now seem to be failing us. With only a few exceptions, national governments badly bungled the response to the COVID-19 pandemic. In the political sphere, democracy itself is under attack. Urged on by then-president Donald Trump peddling the lie of a fraudulent presidential election, a mob of thousands sacked the U.S. Capitol building in an attempt to overturn the results. They waved flags of the Confederacy within the halls of the Capitol. Some of the seditionists called for killing legislators and then–vice president Mike Pence. As one of the original antifascists, Antonio Gramsci, warned from his prison cell in the 1930s, "The crisis consists precisely in the fact that the old is dying and the new cannot be born, in this interregnum a great variety of morbid symptoms appear."[1] Looking ahead into the maw of climate change, capitalism, as it has been practiced, seems to be incompatible both with democracy and with the long-term survival of our species. Faced with the rapid advance of global warming, time is running out.

D.I.Y. Saving Our Planet

Some say we should trust in God or capitalism, with the invisible hand of the marketplace, to deliver our rebirth. Call me skeptical. Certainly, progressive businesses, civil institutions, churches, and governments have critical roles to play, but we can't simply wait around for the United Nations, Joe Biden, the Nature Conservancy, or Greta Thunberg to save us. We as individuals must join together to save ourselves and each other. I think of it as DIY saving our planet. If you're looking for a punchline at the end of this book, that is it.

I look most hopefully to the potential for individuals clustered in small groups to make a difference—to work and innovate our way out of the mess we're in. Think about it. Hope, aspiration, curiosity, invention, art, music, dance, belief in ourselves and others, empathy, sharing, love: these are the human qualities that lifted humanity from the muck of animal history, and they are the core elements with which we can build a new society.

Ask yourself: What's wrong with our world? What world do you want to live in? And how can we get there from here? For many of us, the world we hope for is much different than the one we have today. Also, it's obvious that living a virtuous, sustainable, low-consumption life as an individual won't change the calculus of the Earth system sufficiently to bring peace, reduce suffering, and increase environmental sustainability on a global scale. We need to find each other—people of goodwill and good purpose—and combine forces so we have the power to fundamentally transform our societies. (Please insert here your auditory memory of Patti Smith's anthem "People Have the Power," or Pharrell Williams's "Entrepreneur.")

Pivot Projects provides one model for how this might be done. It's a global experiment made up of people who share a common purpose and who are dedicated to helping people around the world identify paths forward that would make their places more sustainable and resilient. Think of Pivot Projects and other organizations with similar goals as catalysts for change and help givers. But it's really the groups in cities and towns and neighborhoods—the thousand points of light—who, in aggregate, have the potential to recharge the world on a massive scale. Pivot Projects and other groups and institutions can help them—and it seems likely that collaborative intelligence, systems thinking, and artificial intelligence (AI) will be useful tools in their hands.

My involvement with Pivot Projects became a circular journey. I joined the group to chronicle its efforts to take on global-scale problems, but Pivot Projects brought me back home and deposited me on my own doorstep. This happened as I was finishing the first draft of this book. At that point, I began pivoting from being a participant-observer to being mainly just a participant. I joined several of the workstreams and took active part in their discussions. They included Education; Arts and Culture; and Economics, Law, and Politics. At a joint meeting of the Arts and Culture and Education groups, we began talking about the challenge of helping people fully appreciate the relationship between humans and nature. We started talking about rivers. Most people have a river or a stream in their neighborhoods, so why not use rivers as a way of engaging and teaching about the environment?

We humans love our rivers. We love to walk along them, sit beside them, fish in them, swim in them, skate on their frozen surfaces, write songs and poems about them, and drink from their reservoirs. But we don't treat our rivers very well. We use them as dumping grounds for human and industrial waste. Our road salt, lawn fertilizer, and pesticides pollute them. And, unfortunately, when we kill our rivers, we kill ourselves.

Our idea is that we can encourage people to appreciate the planet more, to change behavior and thus address global warming, and to help save our species if we reintroduce them to the rivers in their own neighborhoods. So thus began our River Teachers project. I ran with this idea. Near my apartment in New Haven, Connecticut, is the Mill River. I frequently walk beside it. I have also kayaked and fished in the river. I love the Mill, but I realized that I didn't actually know much about it, so I decided to learn more. I'm a documentary filmmaker, among other pastimes, so I decided to explore the river from its source to its terminus in New Haven Harbor and capture what I saw on video.

Although the Mill River is only thirteen miles long, it tells the whole story of humanity's relationships with rivers. It arises in suburban woods and passes through a rustic state park. It's bottled up to make a reservoir, then snakes gracefully through a beautiful urban park. It is crisscrossed by highways, roads, and railroad tracks. Then it enters an industrial zone where, for more than two centuries, it was the city's outhouse. Then it passes into the harbor and mixes with the Long Island Sound. There's a sizeable oystering industry going on in the harbor and sound, so anything bad that goes into the river goes into the oysters and ultimately into the oyster eaters.

I made a short documentary, *A River Speaks*, which became the first fruit of the River Teachers project. The idea was that it would serve as an inspiration for people and groups around the world. The message: explore your neighborhood's river. Walk along it. Meet people on its banks and talk to them. Tell its story to others. Join with people who share your concerns and help save your river. By learning about our rivers, we learn more about ourselves and our complex relationship with nature. And, maybe, just maybe, we can save ourselves.

One cool thing that happened after I started the river documentary project was I began to meet and interact with many people and groups who were already involved with the Mill River in some way: water scientists, environmentalists, historians, community activists, birders, kayakers, trail builders, artists, educators, and more. Some of these people and groups were already working with each other. I joined several of their organizations. Hopefully, we will get things done together. It's a bit like Pivot Projects, only on a local scale.

Remember that photograph of me in the introduction of this book? It captures me with a blissful expression on my face standing in the headwaters of Colorado's pristine Blue River, which is created largely by snowmelt, and which, a few miles further on from where I stood, becomes a great trout stream. Now picture me again. I'm standing in a murky eddy of the Mill River a half mile from my apartment. The gray silt sucks on my boots. The tide is flowing out. See the determined expression on my face? I want to help save the damned thing. The Mill deserves all the respect and care that we give the Blue.

A Time of Renewed Hope

The morning of January 20, 2021, was a glorious time for many people in the United States and around the world who treasure democracy and want to save the planet. Joe Biden and Kamala Harris were sworn in as the new president and vice president of the United States. The dark age of Trumpism, it seemed, was receding. Biden had promised a total reset on environmental issues and had proposed the most progressive environmental agenda ever by any U.S. president. Biden gave a good speech, but he was eclipsed by the verbal genius of Amanda Gorman, a twenty-two-year-old Black poet who had completed her inaugural poem, "The Hill We Climb," in the aftermath of the right-wing insurrection of January 6, 2021. The poem was a rousing but realistic call to progress and optimism. The last lines went like this: "When day comes, we step out of the shade / aflame and unafraid. / The new dawn blooms as we free it. / For there is always light, / if only we're brave enough to see it / If only we're brave enough to be it." Peter Head was so taken by Gorman's poem that he posted it on the Pivot Projects Slack channel along with a two-word commentary: "That's us."

It seemed suddenly like the wind was at our backs. The United States rejoined the Paris Agreement on Climate Change that day. Within a few days, Larry Fink, chair of BlackRock, the world's largest investment firm with $9 trillion under management, issued a letter calling on corporations to reduce their greenhouse gas emissions by 2050. "I believe that the pandemic has presented such an existential crisis—such a stark reminder of our fragility—that it has driven us to confront the global threat of climate change more forcefully and to consider how, like the pandemic, it will alter our lives," Fink wrote. A few days after that, General Motors announced that it would phase out production of gas-powered vehicles by 2035.

Not surprisingly, the people of the Pivot Projects charged into the new year with renewed energy and purpose. Over the end-of-year holidays, the leaders had conducted a series of planning sessions and completed a membership survey aimed at helping to refine the group's goals, its activities, and the shape of the organization. A new goal was to be fully engaged with ten communities on five continents by the end of the year. They would recruit more women and indigenous people to join the core leadership group. And they would create structures for making work assignments, especially when it came to engaging with communities. Pivot Projects was pivoting, shifting from the internal focus of the first year to an outward focus.

There was good news on the community engagement front as well. The United Nations Development Program (UNDP) made a handshake agreement to form a partnership with Pivot Projects. The two groups would collaborate to offer the Pivot Projects technology platform as a decision-making and knowledge-sharing tool for the UNDP's ninety-two development laboratories around the world. They would start by developing and testing the platform in a handful of places, perhaps Ghana and Pakistan.

Just as the events of early 2021 heartened members of Pivot Projects, the group and its ambitions inspire me and make me hopeful. I feel drawn to the vision of a dynamic, hopeful, evolving world that Damian Costello mapped out with the cellular economy. Society is like a living organism. Within it, groups of people self-organize around a mission, tap into human and computing intelligence, and take on problems that are critical to them and their communities.

I'm drawn to the Two-Loop Theory for transforming organizations and society, which Pivot Projects embraced. It emerged from the fertile minds of C. S. (Buzz) Holling, who applied it to ecology, and of Margaret Wheatley and Deborah Frieze, who applied it to organizational change. In the first loop, ecosystems, organizations, and societies are busy dying, yet if they die too rapidly, before we have a path to new systems and structures, we might descend into chaos. So we have to hospice the best features of today's broken systems. Meanwhile, even in the midst of complex situations, we must imagine a (better) way or ways forward.

But we can't create a sustainable future when we are at war with one another. That's why I'm drawn to the ideas of Lord John Alderdice, a member of the UK House of Lords and participant in Pivot Projects, who was one of the brokers of the Northern Ireland peace agreement. John started his professional life as a psychiatrist but eventually entered politics because he wanted to help address the brokenness of society. Now, in addition to his House of Lords duties, he mediates disputes between groups of people. One of his core insights is that deep conflicts can best be resolved if parties come to the table willing to listen and accept that the person on the other side of a conflict has a legitimate point of view. He also believes that, to make a new and better world, we first have to engage with each other rather than attempt to develop whiz-bang plans on a mountaintop and then try to sell them to people. "Resolving problems is about addressing disturbed historical relationships within communities," John said during one of Pivot Projects' discussions of the nature of power. "You need to develop new structures to address the relationships, but engagement comes first—not the other way around. The new structures will emerge from that."

Emerging from the shock of COVID-19, a shattered world confronts its deficiencies and divisions. The task ahead is to not only to recover from the wounds of our era but also to make society and the natural environment more resilient as we lurch into an uncertain future. It seems to me that this should be entered into with a spirit of reconciliation—to address the polarization in the United States over Trumpism; in the United Kingdom, over Brexit; in the conflict zones of the Middle East, the Pacific Rim, and Africa; and everywhere as we seek to address climate change in ways that are fair and effective. South Africa and Rwanda repaired themselves to some extent through truth-and-reconciliation processes, where former adversaries faced the demons in themselves and one another in efforts to build coherent new societies. It's time for those processes to be launched on a global scale.

In all of these approaches, the first step is to connect and bridge the gaps between us. Isolated, we are weak and broken. Connected, we have power and vitality.

Taking Stock

Pivot Projects explored all of these issues. It brought some of them into sharp relief, but left many others unresolved. Beginning with its original design and going forward as it evolved, the initiative embodied a set of core values aimed at making society and the environment more sustainable and resilient. It brought together a group of people with a common purpose and with a diverse set of skills and points of view. They reasoned together to develop a shared view of how the world works as a system of systems. They tapped into AI to discover ways to make the world work better. And they engaged with communities to help them size up their problems and develop plans for addressing them. Pivot Projects emerged as a whole-Earth hack for the early twenty-first century.

What had Pivot Projects accomplished out of the starting gate? The original plan was to run for nine months and see what could be done. At the beginning of 2021, it was time for a progress check. Peter Head, Colin Harrison, and Rick Robinson had hoped that the twenty workstreams would address major issues in their domains, create system models, and write reports about their findings. Check. The next step was to work with SparkBeyond to test the potential of augmented intelligence and to see if humans collaborating with machines could produce useful and even surprising insights. Because of delays in producing new software features and ingesting the system models, that activity had just begun. Half check. The next step was to engage with communities to help them identify and solve sustainability-related problems. Pivot Projects had handshake agreements with a few places to explore the possibilities or to begin working together. Quarter check.

Then there was the money. The founders had hoped to raise £350,000 to fund operations and salaries but had landed only a tiny fraction of that through a crowd-funding page. They hoped the UK government's AI fund would finally get the money flowing. Head was still running full-tilt at a wall and hoping somebody would open a door and let him in.

On the influence level, they had done better. Head's work on policy briefing papers would be seen by leaders of the Group of Twenty (G20) and the United Nations. They had established a beachhead with Nigel Topping and his team at the twenty-sixth United Nations Climate Change Conference of the Parties (COP26). And between March and December 2020, Peter had presented Pivot Projects' ideas and recommendations in more than twenty virtual conferences attended by more than 800 people.

Perhaps the most promising Pivot Projects offshoot was the Built Environment Pivot Institute (BEPI), the organization that aimed to set new sustainability standards for professionals, universities, and practitioners in the global planning, development, and construction industries. In early 2021, BEPI had honed its ambitions: the top priority was to turn the traditional approach to infrastructure development on its head. Instead of considering the impact of a development project on the environment and society only near the end of the planning process, they wanted it to be addressed from beginning to end. BEPI set out to create a collection of Pivot Projects sustainability performance specifications and make them available to any government in the world who wanted to use them. At the same time, the group felt that it is essential to integrate sustainable planning and building processes and standards with efforts aimed at improving the well-being of individuals within society. In 2015, Wales had passed its Wellbeing of Future Generations Act, which was aimed at improving lives, strengthening communities, and living more sustainably. A number of other countries were considering passing similar legislation. The BEPI team planned on engaging with national governments and urging them to include its standards in legislation they produced.

Looking back on 2020, Pivot Projects' leaders were pleased with the progress that they had made, but they were also frustrated that they hadn't moved farther and faster. "I'm delighted with the overall momentum, the number of people who have joined, and the quality of the output," Peter said. "But I'm really frustrated we haven't been able to do more computing work, and because of that we haven't been able to engage fully with the communities." His new target was to make substantial progress on the AI and engagements in time to present evidence of the model's success to the attendees of COP26 in Glasgow in November 2021. "I think what we're doing will be game changing. We haven't been able to prove it yet, but we will," Peter said.

Colin had been closer than Peter to the alliance with SparkBeyond and to the day-to-day operations of Pivot Projects, attending several workstream meetings each day. He celebrated the fact that this unruly global volunteer organization had gotten off the ground and was still flying. "There is a lesson here about how to create dynamic, transient organizations to solve a specific set of complex problems," Colin said. "It's the kind of outcome we dreamed of in the 1990s as the Internet began to emerge as a real workplace." He had been skeptical at first of SparkBeyond's claims about the capabilities of its computing system. He had wondered if AI would be able to help humans explore the universe of digitally stored knowledge and find relevant, credible, and novel answers to their questions. Now he felt more confident in the outcome. "They have shown it's possible to build insightful, AI-assisted models of real-world systems with relatively modest effort," Colin said. Looking ahead, he thought Pivot Projects demonstrated the kinds of capabilities that will be necessary to query the world's vast and diverse knowledge, navigate complexity, and manage our civilization over the next thirty years. "This will not, by itself, save us from climate-change-induced disasters, but without this kind of capability, we are unlikely to succeed," he said.

To get informed but dispassionate perspectives on Pivot Projects I spoke to several people who were familiar with the initiative but who had not joined the group. Jim Spohrer, a longtime IBMer who had worked with Colin and others on an e-learning campaign, had sat in on some of the Pivot Projects Zoom calls and had given Colin frank feedback on what the group was doing. Spohrer's IBM role at the time was managing strategy for open-source AI projects. He had participated in global collaborations before, collaborations concerning open-source software governance and AI ethics, and believes they served useful purposes. But he believes that AI is overhyped, and it will be years before it will play a powerful role in business and society. Still, he applauded Pivot Projects' mission and strategy. "Some people think that small things lead only to small changes and big things lead to big changes, but I disagree," Spohrer said. "If you get people thinking about systems differently, there's always hope for big change."

Tim Lenton, the professor at University of Exeter who was collaborating with Pivot Projects on education and research at the Exeter Living Laboratory, said that we can't depend on government policy and regulation alone to turn the tables on climate change: we need grassroots efforts to compliment them. "My perception is that, if we're going to have the rate and nature of transformational change we need to decarbonize, we will need networked, bottom-up initiatives that learn by doing and share the learning that's happening around the world," Lenton said. "Pivot Projects speaks to that." By early 2021, the Exeter Living Laboratory was up and running, with a dozen or so graduate students exploring potential projects and checking in at weekly meetings. The group raised money via the university, and Resilience Brokers hired James Green, one of Pivot Projects' young leaders, as its community manager.

Within Pivot Projects, people who invested a lot of time in it looked back on the first months with a mixture of optimism and realism. Jeff Newman, the activist rabbi emeritus, had pulled back from the group when he and his wife were beset with health problems. His enthusiasm and probing questions were missed by others. While he said participating in the group had been a deeply rewarding experience, Jeff questioned how much it could get done. "We know this project won't change the world, but we hope that maybe it will be one small step on the way," he said. "It's satisfying to feel we have been part of something that has helped, just given a little extra help and shape." In contrast, Waël Alafandi, the young Syrian refugee living in France, was inspired by the opportunity to interact with people who shared a common purpose. He believed the group could make a real difference. "I learned that being hopeful about the world hasn't been wrong," he said. "There are plenty of people who really care and are as eager as I am to make the world a better place."

A number of volunteers came away with a feeling that *they* had been changed in notable ways. In some cases, they said, they appreciated learning from people who were different than them or they became more open to new ideas. Several people set off in new career directions or life missions. Deborah Rundlett, the protestant minister in Connecticut, decided to leave her pastoral duties behind and dedicate herself totally to coaching individuals and groups of people that are committed to social change. Damian Costello, the Irish business consultant, had spent the previous decade advising multinational medical device companies on how to fend off innovative upstarts. Now he realized that upsetting the status quo is necessary to foster a healthier society. He took a job with a start-up aiming to disrupt the medical diagnostics field. For Andre Hamm, who had recently graduated from college and was working as a programmer in a Salt Lake City software company, Pivot Projects had made him think differently about his future. He wants a career focused on serving people and the planet. "It has inspired me to be part of the bigger conversation for change," he said. "I want to wake up in the morning and have a mission. I'm here to be of service to the world. I'm here to help."

My Personal Takeaways

As a journalist, I long shied away from injecting myself into the stories I told. In this book project, playing the role of embedded observer, I felt free for the first time to offer my point of view, my personal anecdotes, and my interpretations of events from time to time. Here, with this final chapter, I am turning the camera on myself. That's not only because I have become a full participant in Pivot Projects but also because I believe our personal journeys are essential parts of the roles we play in our communities, in industries, and in society as a whole. Here are my personal takeaways.

We must know ourselves better and be willing to improve ourselves

I have always seen myself as a person of action. I "get shit done." Yet Pivot Projects taught me to slow down a bit, to listen to others with an open mind, and to have the patience to let things develop without so much prodding. I feel this is a valuable lesson not just for me but for two cohorts of which I am a member: the baby boomer generation and the male gender. Deborah Rundlett created a small subgroup within Pivot Projects that focused in part on personal growth. The group used the phrase "let go to let come" as their mantra. They adopted it from Otto Scharmer, a professor at the Massachusetts Institute of Technology (MIT) Sloan School of Management and cofounder of the Presencing Institute, which combines science and personal development to help organizations and communities address challenges and make fundamental changes. To Scharmer, "let go to let come" means we as individuals and groups must let go of our prejudices, our fixed ways of getting things done, and our desire to control situations. We have to be willing to die, metaphorically, to be reborn. The same goes for society. "I am interested in not only the entire transformation of organizations and communities but also the entire civilization and culture that we as human beings collectively enact," Scharmer told an interviewer.[2]

We must have empathy for people with different experiences and beliefs

We also need to feel empathy for the people of the future because the things we do or don't do today will have a profound impact on their lives. You might argue that the importance of empathy goes without saying. It's so obvious. But clearly lack of empathy for others is at the root of many of the problems in our world today. There is no shortcut for cultivating empathy. You can't just read about it or watch documentaries or go to church. You have to meet and interact with people who are very different from you. Zoom makes that a real possibility. I struggle with the challenge of empathizing with people who appear hostile to everything I believe in, for instance, the Moslem jihadist who cuts off the heads of infidels or the Trump supporter who conspires to kidnap and put on trial the governor of Michigan. Yet I have to believe that there is a human being hidden away there whose misery deserves our attention so we can try as a civilization to address it. But empathy is meaningless if it doesn't result in action. Michelle Obama, in her speech at the Democratic National Convention in August 2020, said, "I'm echoing the call of John Lewis, who said, 'When you see something that is not right, you must say something. You must do something.' That is the truest form of empathy, not just feeling but doing, and not just for ourselves or our kids, but for everyone—for all our kids."

We must recognize that we are all in this together

I have long held the view that there are fundamentally two types of human beings: those who believe strongly that we are in this together for the long haul and those who focus primarily on serving their own short-term interests. Obviously, that's too simplistic a view. Each of us is a mix of impulses. But there's a spectrum, and many people stand firmly on one side or the other. I think back on some of the discussions that emerged in several of the Pivot Projects workstreams. There was a recognition that, for much of human history after the hunter-gatherer phase, the great mass of people lived in a state of subservience and subjugation. That began to change with the Reformation and Enlightenment. Gradually, individuals in the West gained freedom and agency. For many, the full flowering of this model took hold in the late twentieth century. Now, we seem to be at a turning point. Will the pull of individual freedom continue to dominate society, or will a new model emerge that is more collectivist and collaborative? Pivot Projects is an example of this model in action—an effort to exploit collaborative intelligence, the soul of the new machine (as I described it in the introduction), for the greater good. I believe that we must embrace the collectivist impulse and run thousands of experiments of this type if we hope to take on the challenges of climate change, inequity, and social disharmony. Interdependence is not a weakness. It is a strength.

I would like to play a role in the big pivot, in reimaging and reinventing the world, but, like many of the Pivot Projects volunteers, I'm an older human, sixty-nine years old at this writing. I believe that it is time for the gray-hairs like me to let go of our grip on the world and hand the reins to younger people, especially young women. It's their world and their future. They have the fresh ideas, the energy, and the stamina to get it done.

Connecting

All-hands meetings of Pivot Projects were scheduled for ninety minutes, but they often lasted nearly two hours. There was so much to pack in, and the participants were in no hurry to take leave of one another. Some people were isolated and hungry for human contact. This activity gave the volunteers hope that they could *do* something. They could contribute. They weren't helpless.

Frequently, all-hands meetings ended on a touching note. People on the Zoom calls stretched to sum things up and to find meaning in what they were doing together. At the end of one such meeting, they discussed the value of collaboration and how modern technologies enabled mind-melds like this one to happen—which would never have been possible before. Peter Williams, the water expert, commented that he was reminded of the brief epigraph in E. M. Forster's novel *Howards End*: "Only connect . . ."[3] "Only connect the prose and the passion, and both will be exalted," Peter said, quoting a line in the book from which the epigraph is taken. Then he extrapolated: "We're connecting different people's proses with different people's passions, different generations' proses with different generations' passions, and different disciplines' proses with different disciplines' passions. We're making all these connections. We're the conduit between all those things. If you can do that, then everything else happens."

* * *

Pivot Projects' connector-in-chief was Alan Dean. He was a longtime friend of Peter Head who ran an education nonprofit in the UK called Burning2Learn. He led the Education workstream and participated in many others. He produced a flood of encouragement and new ideas—and he called "bullshit" when he saw it. Perhaps his most important role was in bringing on and supporting young people. He became a mentor to Tu Anh Ha, the young Vietnamese teacher who had been studying in Europe when COVID hit and was stranded there for more than a year—unable to return home.

Alan died of a heart attack on April 23, 2021, just a few hours after he participated in the weekly Friday all-hands meeting. Word of his death rippled through the organization via Slack and email over the weekend, and on the following Tuesday, during the time slot ordinarily reserved for the weekly Education workstream meeting, the group held a memorial service on Zoom. It was a profoundly sorrowful gathering. Even though most of the people there had never met Alan in person, they felt a deep kinship with him. They shared stories and feelings. James Green played a song on his guitar that he had written after he learned of Alan's death. It sounded a bit like the beautiful and haunting guitar riff in The Beatles' *Blackbird*. Tu Anh described what Alan meant to her: "When I met him, it was a very difficult time in my life. When he came he always spoke with very kind words." She paused, cried for half a minute, and then resumed speaking: "Sometimes I don't feel like I'm a good person, but I want to be the person he said to me. I feel that his words can actually save a person."

In the days after Alan's death, several members of Pivot Projects pledged to redouble their efforts in the group, and a handful joined the core leadership team. James spoke for all of them when he said: "We have to make sure that we're his legacy, especially the young people. We have to keep driving forward."

Still, the future of Pivot Projects was in doubt. Social tides were shifting. The COVID-19 cloud was lifting. People were focusing more on their jobs or were starting new ones. Some of the most avid early participants rarely showed up for meetings anymore. While several of the group's initiatives were progressing, others had languished. A number of the participants had turned their focus on helping out with sustainability projects in their own communities, sort of Pivot Projects Lite—putting the ideas and principles to work on a smaller scale.

In the end, this collective journey and this book aren't about breakthrough technologies or bold new plans for making the world more sustainable, although, those are admirable pursuits. To me, Pivot Projects is, at its essence, about sweet humanity. A group of people, driven by a burning desire to make things better for everybody, struggled and stretched to understand how the world works and to make a difference—to repair Earth and humanity. It was a Sisyphean task, the longest of long shots, yet redeeming.

The journey of Pivot Projects taught many lessons, but none was more important than this: we can only deal with the complexity of the world, of entanglement, of everything being connected to everything else, if we connect to one another, share knowledge and sentience, and collaborate to make to ourselves, each other, and the world around us better.

As E. M. Forster wrote: "Only connect . . ."

Notes

1. The Mission

1. Sarah Kaplan, "50 Years Later, Earth Day's Unsolved Problem: How to Build a More Sustainable World," *Washington Post*, April 21, 2020, https://www.washingtonpost.com/climate-environment/2020/04/21/earth-day-50th-anniversary/.
2. Rebecca Lindsey and LuAnn Dahlman, "Climate Change: Global Temperature," Climate.gov, August 14, 2020, https://www.climate.gov/news-features/understanding-climate/climate-change-global-temperature.
3. "Rise of Carbon Dioxide Unabated," *NOAA Research News*, June 4, 2020, https://research.noaa.gov/article/ArtMID/587/ArticleID/2636/Rise-of-carbon-dioxide-unabated.
4. Jorden Randers and Ulrich Goluke, "An Earth System Model Shows Self-Sustained Melting of Permafrost Even if All Man-Made GHG Emissions Stop in 2020," *Scientific Reports* 10, no. 18456 (2020), https://doi.org/10.1038/s41598-020-75481-z.
5. Mauro Bologna and Gerardo Aquino, "Deforestation and World Population Sustainability: A Quantitative Analysis." *Scientific Reports* 10, no. 7631 (2020), https://doi.org/10.1038/s41598-020-63657-6.
6. Wijnand de Wit, Arianna Freschi, and Emily Trench, "COVID-19: An Urgent Call to Protect People and Nature," *World Wildlife Fund for Nature* (June 2020): 5, https://wwfeu.awsassets.panda.org/downloads/wwf_covid19_urgent_call_to_protect_people_and_nature_1.pdf.
7. Bo Pieter Johannes Andree, "Incidence of Covid-19 and Connections with Air Pollution Exposure: Evidence from the Netherlands," *World Bank Policy Research Working Paper No. 9221* April 27, 2020, https://ssrn.com/abstract=3584842.

8. Jane Goodall, "Jane Goodall: Humanity Is Finished if It Fails to Adapt After COVID-19," *Guardian*, June 3, 2020, https://www.theguardian.com/science/2020/jun/03/jane-goodall-humanity-is-finished-if-it-fails-to-adapt-after-covid-19.
9. Ernesto Londoño and Letícia Casado, "Amazon Deforestation in Brazil Rose Sharply on Bolsonaro's Watch," *New York Times*, November 18, 2019, https://www.nytimes.com/2019/11/18/world/americas/brazil-amazon-deforestation.html.
10. António Guterres, "Secretary-General's Message," *United Nations*, April 22, 2020, https://www.un.org/en/observances/earth-day/message.
11. "Green Stimulus Needed for a Clean Energy Future," *Bloomberg News*, June 9, 2020, https://www.bloomberg.com/features/2020-green-stimulus-clean-energy-future.
12. "The Lockdown Goes Viral," *Economist*, March 20, 2020, https://www.economist.com/graphic-detail/2020/03/20/the-lockdown-goes-viral.
13. Neil M Ferguson, Daniel Laydon, Gemma Nedjati-Gilani, Natsuko Imai, Kylie Ainslie, Marc Baguelin, Sangeeta Bhatia, et al., "Report 9: Impact of Non-Pharmaceutical Interventions (NPIs) to Reduce COVID-19 Mortality and Healthcare Demand," Imperial College COVID-19 Response Team, March 16, 2020. https://www.imperial.ac.uk/media/imperial-college/medicine/sph/ide/gida-fellowships/Imperial-College-COVID19-NPI-modelling-16-03-2020.pdf
14. Peter Senge, "What Is Systems Thinking? Peter Senge Explains Systems Thinking Approach and Principles," *Mutual Responsibility*, http://www.mutualresponsibility.org/science/what-is-systems-thinking-peter-senge-explains-systems-thinking-approach-and-principles.
15. Earth Charter International, "The Earth Charter," https://earthcharter.org/read-the-earth-charter/.

2. The Core Team

1. Peter Head, "Entering the Ecological Age: The Engineer's Role," *Institution of Civil Engineers Brunel Lecture Series* (2008), https://ecosequestrust.org/latest/publications/entering-the-ecological-age-the-engineers-role-peter-head/.
2. "Roadmap 2030: Financing and Implementing the Global Goals in Human Settlements and City-Regions," World Urban Campaign, March 2016, http://www.ecosequestrust.org/roadmap2030.pdf.
3. "World Population Monitoring: Focusing on Population Distribution, Urbanization, Internal Migration and Development," United Nations Economic and Social Council (2008), https://digitallibrary.un.org/record/620038?ln=en.

4. Anthony Leiserowitz, Edward Maibach, Seth Rosenthal, John Kotcher, Parrish Bergquist, Matthew Ballew, Matthew Goldberg, Abel Gustafson and Xinran Wang, "Climate Change and the American Mind: April, 2020," *Yale Program on Climate Change Communication*, May 19, 2020, https://climatecommunication.yale.edu/publications/climate-change-in-the-american-mind-april-2020/.
5. Katie Glueck and Lisa Friedman, "Biden Announces $2 Trillion Climate Plan," *New York Times*, July 14, 2020, https://www.nytimes.com/2020/07/14/us/politics/biden-climate-plan.html.

3. The Scrum

1. "The New Climate Economy: The 2018 Report of the Global Commission on the Economy and Climate," https://newclimateeconomy.report/2018.
2. Joseph E. Stiglitz, "GDP Is the Wrong Tool for Measuring What Matters," *Scientific American* (August 2020), https://www.scientificamerican.com/article/gdp-is-the-wrong-tool-for-measuring-what-matters/.
3. Tega Brain, "The Environment Is Not a System," Research Values 2018, Aarhus University, December 20, 2017, https://researchvalues2018.wordpress.com/2017/12/20/tega-brain-the-environment-is-not-a-system/.
4. George H. W. Bush, "George Bush Inaugural Address," January 20, 1989, https://www.bartleby.com/124/pres63.html.

4. Struggles

1. Guy Lamb, "The Coming Crime Catastrophe," University of Capetown Safety and Violence Initiative, August 5, 2020, http://www.savi.uct.ac.za/news/coming-crime-catastrophe.

6. The Theory of Everything

1. "Coronavirus (COVID-19) Update: FDA Authorizes Monoclonal Antibody for Treatment of COVID-19," November 9, 2020, U.S. Food and Drug Administration, https://www.fda.gov/news-events/press-announcements/coronavirus-covid-19-update-fda-authorizes-monoclonal-antibody-treatment-covid-19.

Profile: Anna Panagiotou

1. David Snowden, "Complex Acts of Knowing," *Journal of Knowledge Management* 6, no. 2 (May 1, 2002), 100–111, https://www.emerald.com/insight/content/doi/10.1108/13673270210424639/full/html.

7. Rethinking Resilience

1. Bo Zhang, "Top 5 Most Expensive Natural Disasters in History," AccuWeather, March 30, 2011, https://web.archive.org/web/20110331005015/http://www.accuweather.com/blogs/news/story/47459/top-5-most-expensive-natural-d.asp.
2. Martin Fackler, "Tsunami Warnings, Written in Stone," *New York Times*, April 20, 2011, https://www.nytimes.com/2011/04/21/world/asia/21stones.html.
3. Michalea D. King, Ian M. Howat, Salvatore G. Candela, Myoung J. Noh, Seongsu Jeong, Brice P. Y. Noel, Michiel R. van den Broeke, Bert Wouters, and Adelaide Negrete, "Dynamic Ice Loss from the Greenland Ice Sheet Driven by Sustained Glacier Retreat," *Communications, Earth & Environment* 1, no. 1 (2020), https://doi.org/10.1038/s43247-020-0001-2.
4. Sarah Evanega, Mark Lynas, Jordan Adams, and Karinne Smolenyak, "Coronavirus Misinformation: Quantifying Sources and Themes in the COVID-19 'Infodemic,'" The Cornell Alliance for Science, Department of Global Development, Cornell University and Cision Global Insights, https://allianceforscience.cornell.edu/wp-content/uploads/2020/09/Evanega-et-al-Coronavirus-misinformationFINAL.pdf.
5. Timothy S. Paul, Sen Pei, and Sasikiran Kandula, "COVID-19 Projections: Delayed Response to Rebound Would Cost Lives," Columbia Mailman School of Public Health, May 21, 2020, https://www.publichealth.columbia.edu/public-health-now/news/covid-19-projections-delayed-response-rebound-would-cost-lives.
6. Dhaval Dave, Andrew I. Friedson, Drew McNichols, and Joseph J. Sabia, "The Contagion Externality of a Superspreading Event: The Sturgis Motorcycle Rally and COVID-19," IZA Institute of Labor Economics, September 2020, http://ftp.iza.org/dp13670.pdf.
7. Peter Head, "GAR Explainer," U.N. Office for Disaster Risk Reduction, July 28, 2020, https://www.youtube.com/watch?v=mUJnwI6SpkE.
8. Stephen Carpenter and Garry Peterson, "Buzz' Holling, 6 December 1930–16 August 2019." *Nature Sustainability* 2 (2019): 997–998. https://doi.org/10.1038

/s41893-019-0425-9; Andrew Nikiforuk, "Buzz Holling's Resilient Universe," *The Tyee*, November 19, 2019, https://thetyee.ca/Analysis/2019/11/19/Buzz-Holling-Resilient-Universe/.

9. Lance Gunderson and C. S. Holling, editors, "Panarchy: Understanding Transformations in Human and Natural Systems," Washington, DC: Island Press, 2009, https://islandpress.org/books/panarchy.
10. C. S. Holling, "C. S. Holling on Dynamic Resilience," Stockholm Resilience Centre, November 2008, https://youtu.be/FhfmaXZPKEY.
11. "History," U.N. Office for Disaster Risk Reduction, https://www.undrr.org/about-undrr/history.
12. "Chart of the Sendai Framework for Disaster Risk Reduction 2015–2030," U.N. Office of Disaster Risk Reduction, https://www.preventionweb.net/files/44983_sendaiframeworkchart.pdf.
13. "The Ten Essentials for Making Cities Resilient," U.N. Office of Disaster Risk Reduction, https://www.unisdr.org/campaign/resilientcities/toolkit/article/the-ten-essentials-for-making-cities-resilient.
14. "Public Health System Resilience Scorecard," U.N. Office of Disaster Risk Reduction, https://www.unisdr.org/campaign/resilientcities/toolkit/article/public-health-system-resilience-scorecard.
15. Scott Kulp and Benjamin Strauss, "New Elevation Data Triple Estimates of Global Vulnerability to Sea-level Rise and Coastal Flooding. *Nature Communications* 10 (2019): 4844. https://doi.org/10.1038/s41467-019-12808-z.
16. "New WWF Report: Three Billion Animals Impacted by Australia's Bushfire Crisis," WWF-Australia, July 28, 2020, https://www.wwf.org.au/news/news/2020/3-billion-animals-impacted-by-australia-bushfire-crisis#.
17. "California Daily Wildfire Update," *Cal Fire*, September 14, 2020, https://www.fire.ca.gov/daily-wildfire-report/.
18. "CoreLogic Estimates $8 Billion to $12 Billion in Insured Losses from Hurricane Laura Wind and Storm Surge," CoreLogic, August 28, 2020, https://www.corelogic.com/news/corelogic-estimates-8-billion-to-12-billion-in-insured-losses-from-hurricane-laura-wind-and-storm-surge.aspx.
19. "Global Warming and Hurricanes," U.S. National Oceanic & Atmospheric Administration, last revised September 23, 2020, https://www.gfdl.noaa.gov/global-warming-and-hurricanes/.

8. Talking to Robots

1. John E. Kelly III and Steve Hamm, *Smart Machines: IBM's Watson and the Era of Cognitive Computing* (New York: Columbia University Press, 2013), 138, 139.

2. Collin G. Harrison, "Chapter One: Augmented Intelligence and Society," Daniel Araya, ed., *Augmented Intelligence and Society, in Augmented Intelligence: Smart Systems and the Future of Work and Learning* (New York: Peter Lang, 2018), 10–39.
3. Na Zhou, Chuan-Tao Zhang, Hong-Yin Lv, Chen-Xing Hao, Tian-Jun Li, Jing-Juan Zhu, Hua Zhu, et al., "Concordance Study Between IBM Watson for Oncology and Clinical Practice for Patients with Cancer in China," *Oncologist* 24, no. 6 (September 4, 2018): 812–819, doi: 10.1634/theoncologist.2018-0255, Epub 2018 Sep 4. PMID: 30181315; PMCID: PMC6656482, https://theoncologist.onlinelibrary.wiley.com/doi/abs/10.1634/theoncologist.2018-0255.
4. Lisa Ehrlinger and Wolfram Wöß, "Towards a Definition of Knowledge Graphs," Institute for Application Oriented Knowledge Processing, Johannes Kepler University, http://ceur-ws.org/Vol-1695/paper4.pdf.
5. "Trustworthy Artificial Intelligence: Rolls-Royce Publishes Pioneering Bias-Control AI Ethics Toolkit," Rolls-Royce press release, December 14, 2020, https://www.rolls-royce.com/media/press-releases/2020/14-12-2020-trustworthy-artificial-intelligence-rolls-royce-publishes-pioneering-bias-control.aspx.
6. David Rolnick, Priya L. Donti, Lynn H. Kaack, Kelly Kochanski, Alexandre Lacoste, Kris Sankaran, Andrew Slavin Ross, et al., "Tackling Climate Change with Machine Learning," Climate Change A.I., https://arxiv.org/pdf/1906.05433.pdf.

9. Points of Light

1. Town of Chapel Hill, "Chapel Hill Mobility and Connectivity Plan," last updated October 28, 2020, https://www.townofchapelhill.org/residents/transportation/bicycle-and-pedestrian/chapel-hill-mobility-and-connectivity-plan.
2. George Day, "Is It Real? Can We Win? Is It Worth Doing? Managing Risk and Reward in an Innovation Portfolio," *Harvard Business Review*, December 2007, https://hbr.org/2007/12/is-it-real-can-we-win-is-it-worth-doing-managing-risk-and-reward-in-an-innovation-portfolio.
3. Jeroen Warner and Arwin Van Buuren, "Implementing Room for the River: Narratives of Success and Failure in Kampen, the Netherlands," *International Review of Administrative Sciences* 77, no. 4 (2011): 779–802, https://www.researchgate.net/publication/258182598_Implementing_Room_for_the_River_Narratives_of_success_and_failure_in_Kampen_the_Netherlands;

Tracy McVeigh, "The Dutch Solution to Floods: Live with Water, Don't Fight It," *Guardian*, February 15, 2014, https://www.theguardian.com/environment/2014/feb/16/flooding-netherlands.
4. Darren Jones and Mel Stride, "The Path to Net Zero," Climate Assembly U.K., https://www.involve.org.uk/sites/default/files/field/attachemnt/Climate%20Assembly%20UK%20Executive%20Summary_0.pdf.

Profile: Gamelihle Sibanda

1. Rajendra B. Magar, Afroz N. Khan, and Abdulrazzak Honnutagi, "Waste Water Treatment Using Water Hyacinth," *32nd Indian Engineering Congress*, The Institution of Engineers (India), 2017, https://www.researchgate.net/publication/323278568_Waste_Water_Treatment_using_Water_Hyacinth.

10. Places

1. "California Statewide Fire Summary September 29, 2020," *Cal Fire*, https://www.fire.ca.gov/daily-wildfire-report/.
2. "Governor Newsom Announces California Will Phase Out Gasoline-Powered Cars & Drastically Reduce Demand for Fossil Fuel in California's Fight Against Climate Change," press released published by the State of California, September 23, 2020, https://www.gov.ca.gov/2020/09/23/governor-newsom-announces-california-will-phase-out-gasoline-powered-cars-drastically-reduce-demand-for-fossil-fuel-in-californias-fight-against-climate-change/.
3. Dave Snowden, "Cynefin as of 1st June 2014," Cynefin Centre, https://commons.wikimedia.org/wiki/File:Cynefin_as_of_1st_June_2014.png.
4. Patrick Greenfield, "World Leaders Pledge to Halt Earth's Destruction Ahead of U.N. Summit," *The Guardian*, September 27, 2020, https://www.theguardian.com/environment/2020/sep/28/world-leaders-pledge-to-halt-earth-destruction-un-summit.
5. Simon Sharpe and Timothy M. Lenton, "Upward-Scaling Tipping Cascades to Meet Climate Goals—Plausible Grounds for Hope," Policy Briefing Note series 2020/01, Global Systems Institute, https://www.exeter.ac.uk/media/universityofexeter/globalsystemsinstitute/documents/202001briefingnote.pdf.
6. "Building a Future City," Future City Glasgow, 2013, https://futurecity.glasgow.gov.uk/reports/12826M_FutureCityGlasgow_Evaluation_Final_v10.0.pdf.

7. "Digital Glasgow: 2020 Review," Digital Glasgow Board, 2020, https://www.glasgow.gov.uk/councillorsandcommittees/viewSelectedDocument.asp?c=P62AFQDNDX0G2U812U.
8. Jorgen Randers and Ulrich Goluke, "An Earth System Model Shows Self-Sustained Melting of Permafrost Even if All Man-Made GHG Emissions Stop in 2020," *Nature Scientific Reports* 10, no. 18456 (2020), https://doi.org/10.1038/s41598-020-75481-z.

Profile: Paola Bay

1. Mamo Dwawiku Izquierdo, "Don't Say They Didn't Tell Us," translated by Amanda Bernal-Carlo, The Great Balance, March 27, 2020, https://www.thegreatbalance.org/2020/03/27/dont-say-they-didnt-tell-us/.

11. Bright Ideas

1. Peter Head, Ryan Bartless, Rowan Palmer, Steven Crosskey, and Anuj Malhotra, "Policies and Implementation Guidelines for Data-Driven, Integrated, Risk-Based Planning of Sustainable Infrastructure," Saudi T20, https://t20saudiarabia.org.sa/en/briefs/Pages/Policy-Brief.aspx?pb=TF3_PB4.
2. Brandon Graver, Kevin Zhang, and Dan Rutherford, "CO_2 Emissions from Commercial Aviation, 2018," International Council on Clean Transportation, https://theicct.org/sites/default/files/publications/ICCT_CO2-commercl-aviation-2018_20190918.pdf.
3. Donella H. Meadows, Dennis L. Meadows, Jorgen Randers, and William W. Behrens III, *The Limits to Growth* (Universe Books, 1972), http://www.donellameadows.org/wp-content/userfiles/Limits-to-Growth-digital-scan-version.pdf.
4. Helen Mountford, Jan Corfee-Morlot, Molly McGrego, et al., *The 2018 Report of the Global Commission on the Economy and Climate*, New Climate Economy, https://newclimateeconomy.report//2018.
5. Daniel W. O'Neill, Andres L Fanning, William F. Lamb, and Julia K. Steinberger, "A Good Life for All Within Planetary Boundaries." *Nature Sustainability* 1 (2018): 88–95, https://doi.org/10.1038/s41893-018-0021-4.
6. Thomas Piketty, *Capital in the Twenty-First Century* (Boston, MA: Harvard University Press, 2013).

7. Joel Millward-Hopkins, Julia K. Steinberger, Narasimha D. Rao, and Yannick Oswald, "Providing Decent Living with Minimum Energy: A Global Scenario," *Global Environmental Change* 65 (2020): 102–168, http://pure.iiasa.ac.at/id/eprint/16764/1/1-s2.0-S0959378020307512-main.pdf.
8. "C40 Mayors' Agenda for a Green and Just Recovery," The C40 Knowledge Hub, July 2020, https://www.c40knowledgehub.org/s/article/C40-Mayors-Agenda-for-a-Green-and-Just-Recovery?language=en_US.
9. "Issues Brief: Forests and Climate Change," International Union for Conservation of Nature, November 2017, https://www.iucn.org/sites/dev/files/forests_and_climate_change_issues_brief.pdf.
10. U. Ilstedt, A. Bargués Tobella, H. R. Bazié, J. Bayala, E. Verbeeten, G. Nyberg, J. Sanou, et al., "Intermediate Tree Cover Can Maximize Groundwater Recharge in the Seasonally Dry Tropics," Center for International Forestry Research, *Scientific Reports* 6 (2016): 219–230, https://doi.org/10.1038/srep21930.
11. Earth Charter Commission, "The Earth Charter," https://earthcharter.org/wp-content/uploads/2020/03/echarter_english.pdf.

12. Connecting

1. Antonio Gramsci, *Selections from the Prison Notebooks* (New York: International Publishers, 1971), 276.
2. Otto Scharmer: "Theory U: Leading from the Future as It Emerges", *Integral Leadership Review* (August–November 2013), http://integralleadershipreview.com/10916-otto-scharmer-theory-u-leading-future-emerges/.
3. E. M. Forster, *Howard's End* (New York: Alfred A. Knopf, 1921), title page.